Computational Probability

Academic Press Rapid Manuscript Reproduction

The Proceedings of the
Actuarial Research Conference on Computational Probability
Held at Brown University, Providence, Rhode Island
on August 28–30, 1975
Sponsored by the
Committee on Research of the Society of Actuaries
and the Department of Applied Mathematics of Brown University

Computational Probability

Edited by

P. M. Kahn
San Francisco, California

1980

Academic Press
A Subsidiary of Harcourt Brace Jovanovich, Publishers
New York London Toronto Sydney San Francisco

ACADEMIC PRESS, INC.
111 Fifth Avenue, New York, New York 10003

United Kingdom Edition published by
ACADEMIC PRESS, INC. (LONDON) LTD.
24/28 Oval Road, London NW1 7DX

Library of Congress Cataloging in Publication Data
Main entry under title:

Computational probability.

Proceedings of a conference held at Brown University,
Aug. 28-30, 1975.
1. Insurance—Mathematics—Congresses. 2. Insurance
—Statistical methods—Congresses. 3. Probabilities—
Congresses. I. Kahn, Paul Markham.
HG8781.C63 368'.001'5192 80-15014
ISBN 0-12-394680-8

PRINTED IN THE UNITED STATES OF AMERICA

80 81 82 83 9 8 7 6 5 4 3 2 1

This volume is dedicated to the late

DAVID GARRICK HALMSTAD

Cochairman of this Conference
and Secretary to the
Committee on Research of the Society of Actuaries.

Contents

Contributors

Numbers in parentheses indicate the pages on which the authors' contributions begin.

John A. Beekman (287), Department of Mathematical Sciences, Ball State University, Muncie, Indiana

Gottfried Berger (185), Cologne Life Reinsurance Company, P.O. Box 300, Stamford, Connecticut

Harald Bohman (249, 257), Skandia Insurance Company, Ltd., Box S-103 60, Stockholm, Sweden

Phelim Paul Boyle (91), Faculty of Commerce and Business Administration, University of British Columbia, Vancouver, British Columbia, Canada

J. R. Brzezinski (325), Life Insurance Marketing and Research Association, 170 Sigourney Street, Hartford, Connecticut

Lena Chang (231), Actuarial and Insurance Consultant, Chang & Cummings, 6 Beacon Street, Boston, Massachusetts

Clinton P. Fuelling (287), Department of Mathematical Sciences, Ball State University, Muncie, Indiana

Hans U. Gerber (261), Department of Mathematics, The University of Michigan, Ann Arbor, Michigan

Ulf Grenander (1), Division of Applied Mathematics, Brown University, Providence, Rhode Island

T. N. E. Greville (173), National Center for Health Statistics, Rockville, Maryland

Donald A. Jones (261), Department of Mathematics, The University of Michigan, Ann Arbor, Michigan

Nathan Keyfitz (173), Department of Sociology, Harvard University, Cambridge, Massachusetts

Kenneth M. Levine (131), Insurance Company of North America, 1600 Arch Street, Philadelphia, Pennsylvania

Peter Masters (131), Insurance Services Office, 160 Water Street, New York, New York

Donald E. McClure (311), Division of Applied Mathematics, Brown University, Providence, Rhode Island

Donald R. McNeil (39), School of Economics and Financial Management, Macquarie University, North Ryde, 2113 New South Wales, Australia

Marcel F. Neuts (11), Department of Mathematical Sciences, University of Delaware, Newark, Delaware

Donald L. Orth (155), IBM Corporation, 1700 Market Street, Philadelphia, Pennsylvania

Siegfried Schach (39), Department of Statistics, The Johns Hopkins University, Baltimore, Maryland

Robert Sedgewick (101), Department of Computer Science, Brown University, Providence, Rhode Island

Arnold F. Shapiro (71), Department of Business Administration, The Pennsylvania State University, University Park, Pennsylvania

Richard Vitale (303), Department of Mathematics, Claremont Graduate School, Claremont, California

Nils Wikstad (337), Försäkringstekniska Forskningsnämnden, Stockholm, Sweden

Preface

This volume is a collection of papers presented at the Actuarial Research Conference on Computational Probability and related topics held at Brown University, August 28–30, 1975. The conference was sponsored jointly by the Committee on Research of the Society of Actuaries and the Department of Applied Mathematics of Brown University. The cochairmen of the conference were Walter Freiberger and David G. Halmstad.

The principal objective of the conference was to explore the development of computational techniques in probability and statistics and their application to problems in insurance. Papers may be grouped under the six general topics: computational probability, computational statistics, computational risk theory, analysis of algorithms, numerical methods, and notation and computation. Applications covered both life and nonlife insurance.

We wish particularly to express our gratitude to Brown University for the use of facilities in Barus-Holley Hall, and particularly to Merton P. Stoltz, Provost of the University, for his warm welcome. The editor wishes to record his special thanks to Mrs. Katherine MacDougall who typed the entire manuscript for publication.

SOME IDEAS IN COMPUTATIONAL PROBABILITY

Ulf Grenander

Division of Applied Mathematics
Brown University
Providence, Rhode Island

I would like to start with a question: what has the computer meant for mathematics, how has it influenced the way mathematicians work? I am thinking of mathematics itself, the theory, not how partial differential equations or large statistical data bases are treated numerically.

After all, some of the pioneers in computers were mathematicians, von Neumann, Goldstine and others, and it would be reasonable to expect that mathematicians would be the first to use computers. In spite of this, I doubt that many professional mathematicians would admit that computers have played an important role in contemporary mathematics.

When computers began to be available to the academic community their importance became obvious to workers in many fields: physics, biology, medicine, geology, actually all the physical sciences. Examples may be found also in the social sciences, such as statistical data analysis and even in humanities more recently, such as the "cliometricians" in history.

ISBN 0-12-394680-8

In mathematics we did not have the same sort of enthusiastic acceptance of the computer as a new research tool, except for isolated cases such as Lehmer's work in number theory. Why is this?

Let us narrow down the question to the branch of mathematics that we should be concerned with at this conference: probability. What can computers do for the probabilist?

It is important to realize the qualitative difference between mathematics and the physical sciences. Chemistry or astronomy deals with certain aspects of the physical world - mathematics, probability say, deals with an abstract world constructed by man. The probabilist deals with logical structures, not with lots of data (although his cousin, the statistician, may do this) so that he will usually not be confronted with massive number crunching.

The role of the computer will be of a different kind in our context, and there were people who realized this quite early. In 1968 John Tukey and Frederick Mosteller stated about the influence of the computer:

> Ideally we should use the computer the way one uses paper and pencil: in short spurts, each use teaching us - numberically, algebraically, or graphically - a bit of what the next use should be. As we develop the easy back-and-forth interaction between man and computer today being promised by time-sharing systems and remote consoles, we shall move much closer to this ideal.

Today, in 1975, we are indeed closer to this ideal. Although the mode of operation described by Tukey and Mosteller has not yet been widely adopted, it has become clear to a widening circle of practitioners of mathematical statistics that we can and must use the new possibilities of interactive computer.

At Brown University we initiated a project in 1967 called "Computational probability." Its purpose was to explore the consequences for this area of applied mathematics of technological advances, not just the increased speed and storage capacity of the computer, but also from time-sharing systems, new programming tools, and graphical displays.

During the academic year 1968-69 a graduate course was offered in computational probability to test some of the teaching material that had been developed. At the same time we let the students investigate a number of research problems, some of which led to M.A. or Ph.D. theses. We encouraged an experimental approach, using the computer as the physicist uses his laboratory.

This course has afterwards been modified and repeated on different levels. Some of the material can be found in the book by Walter Freiberger and myself.

Let us start by describing how we tried to implement the program insofar as teaching was concerned. Examples were presented, either in lectures together with the computational results or, more often, handed out to the students with some hints about possible ways of approaching the problem and asking them to do what they could to solve it. The examples presented were chosen because they would give the student a feeling for the degree of mathematical complexity that often arises in real-life applications as contrasted with the well-behaved textbook examples tailored to fit existing theory. A certain amount of time has to be spent presenting the subject matter background, but this is unavoidable if one wishes to describe and criticize the crucial relation between model and reality. Otherwise one could easily gloss over someof the messy details in the problem and smooth out realistic features of it. All models represent idealizations to some degree and a difficult phase in the model-building process is the choice of what features should be accounted for in the model and what one feels can be left out safely. How to strike this balance should be discussed in class.

Among examples of this type let us briefly mention a few. Many of the fundamental limit theorems were studied both by simulations and by direct computing. Simulation is usually easier but it has the drawback that the accuracy is low for reasonable computing cost. Direct computation is harder and may involve some hard numerical analysis of the algorithms involved.

Particular attention was paid to situations where the limit theorems did not apply, or at least the convergence would be so slow as to make them practically meaningless. The laws of large numbers, the ergodic theorem for stationary stochastic processes, and the central limit theorems were among the ones studied in this way.

We tried to link up the limit theorem with the behavior of the stochastic processes associated with the partial sums. It was noticed early in the project that the didactic value increased dramatically when the results were displayed graphically. In the beginning this was done by simple plots produced by the terminal typewriter; later we experimented with plotters and cathode-ray tubes. More about this later.

The mathematical experiments and the phenomena displayed were later put into the right perspective by discussing the proof of the theorem in question. In this way the student got into the habit of correlating the results of the mathematical experiments with theory and using the computer as a tool to be combined with deduction, not to replace it.

In these instances the underlying theory was known but in many other cases this was not so: either the theory had not yet been developed and we tried to do it, or the model was so complicated that a purely deductive approach seemed hopeless.

It is easiest to illustrate how this was done by discussing some examples. You will hear about several others in Dr. Vitale's paper. Let me start with a recent problem treated in detail in Jack Silverstein's Ph.D. thesis.

Consider the following mathematical formalization of a neurophysiological network. We have n input nodes and n output nodes, with n very large. The strength of the connection from input node i to output node j is some stochastic variable v_{ij}

which are all assumed to be i.i.d. In our linear model, if x is the input vector $x = \{x_i;\ i=1,2,\ldots,n\}$ and y is the output vector $y = \{y_j;\ j=1,2,\ldots,n\}$ we have

$$y = Vx;\ V = \{v_{ij};\ i,j=1,2,\ldots,n\} \tag{1}$$

Hence the signal power of the output is

$$\|y\|^2 = y^T y = x^T V^T V x = x^T W x \tag{2}$$

and the response to various stimuli x is therefore expressed in terms of the W matrix.

The first problem for this model is to characterize W in probabilistic terms for large n, n, in other words to *prove probabilistic limit theorems for stochastic matrices*. Note that the W-entries are not i.i.d. which makes this a hard analytical problem.

I had conjectured that W, after an appropriate normalization, would obey what one could call a *law of large numbers for stochastic matrices* and converge in probability to a constant matrix, in this case the identity matrix.

To try this conjecture we went ahead in the spirit of computational probability, using the machine as our laboratory, and did some experimental mathematics. Let us generate a W-matrix and see if it indeed is close to I. Now W is symmetric and non-negative definite so that its spectrum, the eigenvalue distribution, is on the non-negative part of the real line. If the conjecture is correct this distribution should be concentrated around 1, as in Figure 1.

We were quite surprised, and at first disappointed, in finding something quite different, see Figure 2.

This experiment was repeated in many variations, all the time with results similar to Figure 2, and we were finally forced to look for other conjectures. Exponential distributions perhaps? Well, in most of the experiments the value $\lambda = 4$ seemed to be a sharp cutoff point, and also at $\lambda = 0$ we seemed to be getting

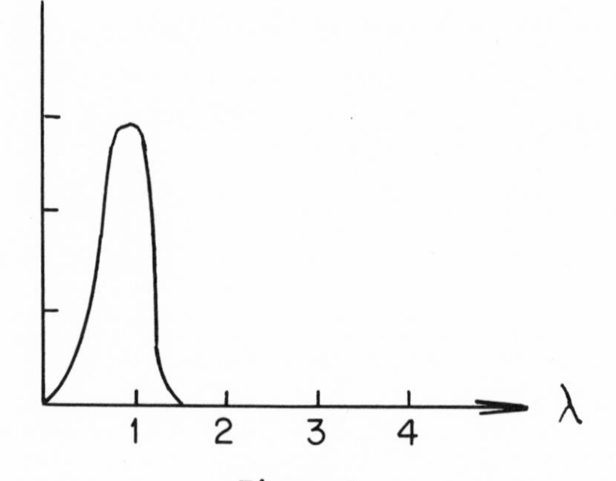

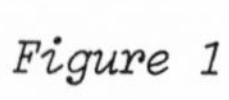
Figure 1

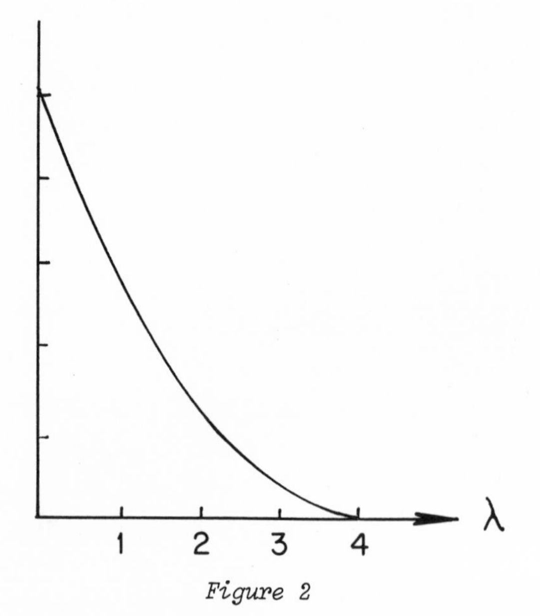

Figure 2

very large values.

We therefore decided to look for other types of distributions. I should not go into details here (see Silverstein's thesis), but we finally succeeded in proving a limit theorem with a universal limit distribution. The proof was long and difficult, but the point is that the computational experiment helped us choose between alternative conjectures and certainly helped us to get to the correct one faster than otherwise would have been possible.

This example involves some sophisticated analysis, but the same experimental approach can be used in more elementary cases. Here is a simple case. I was studying a maximum likelihood estimation of a parameter that was known to be in a certain interval. There were heuristic reasons to believe that the value of the estimate would always be at the boundary of the interval.

The form of the likelihood function was in terms of the usual elementary functions, a fairly long expression. One could no doubt find out analytically if the conjecture was true, it is the sort of thing a student with one year of calculus should be able to do in an afternoon or a day perhaps. Instead I did a simple computation experiment and I found almost immediately that there were cases when the maximum was *not* attained at the boundary. The programming and execution, done interactively took half an hour with a cost of about 20 cents!

Now let us turn to a much more complex type of problem arising in decision making under uncertainty. There is a large body of knowledge accumulated on how to do this rationally. One approach clear to this audience comes from insurance: the collective theory of risk. Another from economics: theory of games.

Both are beautiful theories but severely limited when applied to real life situations. We were interested in more diffuse situations, when the information about the risk structure was less well known, when competition plays a role, when good will and public relations cannot be neglected, when we decide not once and for all but repeatedly, and many other similar complications were included.

The problem was the following. In such situations, is it possible to describe good, if not optimal, behavior in precise terms? "Precise" meaning that they can be implemented in an algorithm, and "good" meaning that if the algorithm is compared in performance to heuristic human behavior it should come out about as well.

Obviously this cannot be solved by completely analytical methods, and we performed a sequence of computational experiments involving various algorithms as well as many human subjects. The results can be found in the Ph.D. thesis by Patrick Bourke and a small number of copies of the thesis are available here.

It will interest the audience that the setup chosen was that of insurance, but I had better emphasize that we were not aiming for a realistic description of any particular insurance activity: the values of the parameters have nothing to do with real risk parameters, the conceptual structure is only superficially similar to an insurance business. The reason for choosing the format of an insurance business was just my old interest in actuarial mathematics; one could just as well have picked banking or some military situation.

I'll mention briefly some of the options open to the players.

1) Buy information, research and development, to counteract the lack of precise knowledge of risk structure.

2) Invest capital in different ways, short range-long term.

3) Spend on public relations, sales promotion, advertising to attract customers.

4) Liquidate assets.

5) Change rates successively as time goes on.

In some versions we have also included reinsurance, restrictions due to a regulatory agency, inflation and so on. The game has some elements in common with management games but this is a mathematical game, quite different in purpose.

After playing the game with human players, we tried out many "algorithmic players" against "human players". There is a good deal of random noise involved so that the experiment has to be replicated carefully. Also the performance of the human players varied a lot. Brown University is fortunate in having some very clever undergraduates and some of them performed exceedingly well compared to others.

This led us to discard many suggested algorithms, and we finally concentrated on six. They are now "tuned" by adjusting parameters and using response surface techniques.

We learnt two things from this. First, the immediate goal, the answer to our question, turned out to be yes: it is indeed possible in situations of high complexity to construct algorithms that perform approximately as well as human intuition in an average sense. It was not an unqualified success however: the algorithm we had expected most from did not perform so well. The explanation came much later and involved an algorithmic error difficult to spot.

Second, and this is my main point here, we encountered a large project in experimental mathematics, where the programming had to be changed again and again as we gained experience about the way human players could be expected to react in different situations. The total programming effort from the beginning of the game was of the order of a couple of thousand APL-statements which would correspond to many thousands of, say, FORTRAN statements. The programming effort was therefore not trivial. It easily happens in a case like this that the programming difficulties take up so much time that the emphasis is shifted from the conceptual and analytical side, where it should be, to programming, which should be only a tool. This was avoided by the ability we had to do this interactively and in a programming language suitable for mathematics. We were lucky to have access to a first rate interactive system at Brown University; without it I doubt that we would have had the tenacity to persevere during all the difficulties we encountered during the experiments.

I think I can speak for all the members, past and present, in the group working on the Computational Probability Project at Brown University, that our way of working has been profoundly affected by what we learnt from the project, and indeed for the better.

REFERENCES

Bourke, P. (1974). The design and improvement of heuristics for an insurance management game, Div. of Applied Mathematics, Brown University.

Freiberger, W. and Grenander, U. (1971). A Short Course in Computational Probability and Statistics, Springer-Verlag, New York.

Grenander, U. (1973). Computational probability and statistics, *SIAM Review, 15*, pp. 134-191.

COMPUTATIONAL PROBLEMS RELATED TO THE GALTON-WATSON PROCESS

Marcel F. Neuts[1]

Department of Statistics[2]
Purdue University
Lafayette, Indiana

SUMMARY

An algorithm to compute the probability distributions of the successive generation sizes in a Galton-Watson process is presented. The distribution of the number of offspring of each individual is assumed to be of phase type. A probability distribution is of phase type if it can be identified as the distribution of the time till absorption in an absorbing finite Markov chain with appropriate initial conditions. A detailed analysis of the error due to truncation is given, as well as an application in a problem related to the M/G/1 queue.

[1]*Research sponsored by the Air Force Office of Scientific Research, Air Force Systems Command, USAF, under Grant No. AFOSR-72-2350B. The United States Government is authorized to reproduce and distribute reprints for governmental purposes notwithstanding any copyright notation hereon.*

[2]*Present address: Department of Mathematical Sciences, University of Delaware, Newark, Delaware.*

ISBN 0-12-394680-8

A second algorithm deals with the distribution of the maximum generation size before extinction. Several theorems on probability distributions of phase type are proved.

I. THE PROBABILITY DISTRIBUTIONS OF PHASE TYPE

In computational probability problems it is frequently desirable to have available a versatile class of distributions which may be concisely represented, which satisfy simple recurrence relations, in addition to having algorithmically useful properties under the operations of convolution and mixing. The probability distributions of phase type, which were discussed in [8], have many such properties in addition to the ones proved here. In the present paper, the computational advantages of these distributions in the study of the Galton-Watson process will be examined.

Let $\underline{\alpha}$ denote an m-vector of probabilities and $\underline{e}$ an m-vector with $e_i = 1$, for $i = 1,\ldots,m$. We define α_{m+1} by $\alpha_{m+1} = 1-\underline{\alpha}\,\underline{e}$, and assume that $\alpha_{m+1} \geq 0$. The matrix T of order m is substochastic and such that I-T is nonsingular. The vector $\underline{T}^o$ is defined by $\underline{T}^o = \underline{e} - T\underline{e}$. The matrix T^o of order m has m identical columns given by $\underline{T}^o$. The matrix A is $\mathrm{diag}(\alpha_1,\alpha_2,\ldots,\alpha_m)$.

A probability density $\{p_k\}$ on the nonnegative integers is *of phase type* and has representation $(\underline{\alpha},T)$ if and only if

$$p_0 = \alpha_{m+1}, \qquad p_k = \underline{\alpha}\, T^{k-1}\underline{T}^o, \quad \text{for } k \geq 1. \tag{1}$$

It is shown in [8] that all probability densities on a finite number of nonnegative integers and all generalized negative binomial densities are of phase type. It is further shown that all right shifts, all finite mixtures and all finite convolution products of distributions of phase type are themselves of phase type and appropriate representations for these are constructed.

It is easy to see that $\{p_k\}$ is the probability density of the time till absorption in the (m+1)-state Markov chain with transition probability matrix

$$P = \begin{pmatrix} T & \underline{T}^o \\ \underline{0} & 1 \end{pmatrix}, \tag{2}$$

and initial probability vector $(\underline{\alpha}, \alpha_{m+1})$. The condition that $I - T$ is nonsingular guarantees that absorption into the state $m + 1$ is certain for any initial probability vector. For purposes of representation of the density $\{p_k\}$ we may further assume without loss of generality that the stochastic matrix $Q = T + (1-\alpha_{m+1})^{-1}T^oA$, is irreducible.

There is a completely parallel development for probability distributions of phase type on the nonnegative real line. In this case we consider an (m+1)-state Markov chain in continuous time with infinitesimal generator

$$B = \begin{pmatrix} T & \underline{T}^o \\ \underline{0} & 0 \end{pmatrix}. \tag{3}$$

The matrix T now has negative diagonal elements and nonnegative off-diagonal elements. The matrix T is nonsingular and $T\underline{e} + \underline{T}^o = \underline{0}$. The probability distribution of the time till absorption into the state $m + 1$ with initial probability vector $(\underline{\alpha}, \alpha_{m+1})$ has a jump of α_{m+1} at zero and a density component given by

$$\phi(u) = \underline{\alpha} \exp(uT)\underline{T}^o, \qquad \text{for } u > 0. \tag{4}$$

Phase distributions on the positive real line play only an incidental role in this paper and except where indicated, we shall consider only the discrete case.

Countable mixtures of probability distributions of phase type are generally not of phase type, but the following positive result is of some interest in the Galton-Watson process.

Theorem 1. Let $\{p_k\}$ be a density of phase type with representation $(\underline{\alpha},T)$. Let $\{s_k\}$ be of phase type with representation $(\underline{\beta},S)$, where the order n of the matrix S may be different from that of the matrix T. If P(z) denotes the probability generating function of the density $\{p_k\}$, then the mixture $\{v_k\}$ of the successive convolutions of $\{p_k\}$, whose probability generating function V(z) is given by

$$V(z) = s_0 + \sum_{\nu=1}^{\infty} s_\nu P^\nu(z), \tag{5}$$

is itself of phase type. A representation $(\underline{\gamma},K)$ with K of order mn, of the density $\{v_k\}$ is given by

$$K = T \otimes I_n + T^o A \otimes (I_n - \alpha_{m+1}S)^{-1}S, \tag{6}$$

where $\otimes$ denotes the Kronecker product of matrices, and by the mn-vector $\underline{\gamma}$, given by

$$\underline{\gamma} = \underline{\alpha} \otimes \underline{\beta}(I_n - \alpha_{m+1}S)^{-1}. \tag{7}$$

The corresponding vector $\underline{K}^o$ is given by

$$\underline{K}^o = \underline{T}^o \otimes (I_n - \alpha_{m+1}S)^{-1}\underline{S}^o, \tag{8}$$

and γ_{mn+1} is given by

$$\gamma_{mn+1} = \beta_{n+1} + \alpha_{m+1}\underline{\beta}(I_n - \alpha_{m+1}S)^{-1}\underline{S}^o. \tag{9}$$

Proof. The probability generating functions P(z) and S(z) of $\{p_k\}$ and $\{s_k\}$ are given respectively by

$$\begin{aligned} P(z) &= \alpha_{m+1} + z\underline{\alpha}(I_m - zT)^{-1}\underline{T}^o, \\ S(z) &= \beta_{n+1} + z\underline{\beta}(I_n - zS)^{-1}\underline{S}^o, \end{aligned} \tag{10}$$

and hence

$$V(z) = \beta_{n+1} + \sum_{\nu=1}^{\infty} [\alpha_{m+1} + z\underline{\alpha}(I_m - zT)^{-1}\underline{T}^o]^\nu \underline{\beta} S^{\nu-1}\underline{S}^o$$

$$= \beta_{n+1} + \alpha_{m+1}\underline{\beta}(I_n - \alpha_{m+1}S)^{-1}\underline{S}^o \tag{11}$$

$$+ \sum_{\nu=0}^{\infty} \sum_{j=0}^{\nu} \binom{\nu+1}{j} \alpha_{m+1}^j [z\underline{\alpha}(I_m - zT)^{-1}\underline{T}^o]^{\nu-j+1} \underline{\beta} S^\nu \underline{S}^o.$$

The first two terms yield γ_{mn+1}. It therefore suffices to show that

$$z\underline{\gamma}(I_{mn} - zK)^{-1}\underline{K}^o$$
$$= \sum_{\nu=0}^{\infty} \sum_{j=0}^{\nu} \binom{\nu+1}{j} \alpha_{m+1}^j [z\underline{\alpha}(I_m - zT)^{-1}\underline{T}^o]^{\nu-j+1} \underline{\beta} S^\nu S^o. \tag{12}$$

Observing that

$$[z\underline{\alpha}(I_m - zT)^{-1}\underline{T}^o]^{t+1} = z\underline{\alpha}[z(I_m - zT)^{-1}T^oA]^t(I_m - zT)^{-1}\underline{T}^o, \tag{13}$$

for $t \geq 0$, and interchanging the order of summation, we write the right hand sum as

$$\sum_{j=0}^{\infty} \sum_{t=0}^{\infty} \binom{j+t+1}{j} \alpha_{m+1}^j z\underline{\alpha}[z(I_m - zT)^{-1}T^oA]^t(I_m - zT)^{-1}\underline{T}^o\underline{\beta} S^{j+t}\underline{S}^o. \tag{14}$$

Since

$$\sum_{j=0}^{\infty} \binom{j+t+1}{j} \alpha_{m+1}^j S^j = [I_n - \alpha_{m+1}S]^{-t-2}, \tag{15}$$

the sum reduces to

$$\sum_{t=0}^{\infty} z\underline{\alpha}[z(I_m - zT)^{-1}T^oA]^t(I_m - zT)^{-1}\underline{T}^o\cdot$$

$$\cdot\underline{\beta}[I_n - \alpha_{m+1}S]^{-1}\{[I_n - \alpha_{m+1}S]^{-1}S\}^t[I_n - \alpha_{m+1}S]^{-1}\underline{S}^o$$

$$= z\{\underline{\alpha} \otimes \underline{\beta}[I_n - \alpha_{m+1}S]^{-1}\} \cdot \sum_{t=0}^{\infty} \{z(I_m - zT)^{-1}T^oA \otimes (I_n - \alpha_{m+1}S)^{-1}S\}^t\cdot$$

$$\cdot \{(I_m - zT)^{-1} \otimes I_n\}\{\underline{T}^o \otimes [I_n - \alpha_{m+1}S]^{-1}\underline{S}^o\}, \tag{16}$$

by repeated application of the property

$$(L \otimes M)(K \otimes N) = LK \otimes MN, \tag{17}$$

of the Kronecker product. Finally

$$\sum_{t=0}^{\infty} \{z(I_m - zT)^{-1}T^oA \otimes (I_n - \alpha_{m+1}S)^{-1}S\}^t\{(I_m - zT)^{-1} \otimes I_n\}$$

$$= \{I_{mn} - z(I_m - zT)^{-1}T^oA \otimes (I_n - \alpha_{m+1}S)^{-1}S\}^{-1}\{(I_m - zT)^{-1} \otimes I_n\}$$

$$= \{(I_m - zT) \otimes I_n - zT^oA \otimes (I_n - \alpha_{m+1}S)^{-1}S\}^{-1}$$

$$= [I_{mn} - zT \otimes I_n - zT^oA \otimes (I_n - \alpha_{m+1}S)^{-1}S]^{-1}$$

$$= [I_{mn} - zK]^{-1}. \tag{18}$$

We now turn to the discussion of the existence of the matrix inverses, used in the preceding manipulations. Since $I - T$ and $I - S$ are nonsingular, and since the matrices T and S are nonnegative, the inverses $[I_n - \alpha_{m+1}S]^{-1}$ and $[I_m - zT]^{-1}$, $|z| \leq 1$, exist. This implies the validity of the series representation in formula (15). The validity of the matrix series in formula (18) is shown if the spectral radius of the nonnegative matrix $(I_m - T)^{-1}T^oA \otimes (I_n - \alpha_{m+1}S)^{-1}S$ is less than one. Since

$$(I_m - T)^{-1}T^oA\underline{e} = (1 - \alpha_{m+1})\underline{e}, \tag{19}$$

the matrix $(1 - \alpha_{m+1})^{-1}(I_m - T)^{-1}T^oA$ is substochastic. The matrix $(1 - \alpha_{m+1})(I_n - \alpha_{m+1}S)^{-1}$ is clearly also substochastic, so that the spectral radius of $(1 - \alpha_{m+1})(I_n - \alpha_{m+1}S)^{-1}S$ is strictly less than one. The mn eigenvalues of the Kronecker product $(I_m - T)^{-1}T^oA \otimes (I_n - \alpha_{m+1}S)^{-1}S$ are the products of the eigenvalues of the matrices $(1 - \alpha_{m+1})^{-1}(I_m - T)^{-1}T^oA$ and $(1 - \alpha_{m+1})(I_n - \alpha_{m+1}S)^{-1}S$. All these products are less than one in modulus, which proves the desired result. The properties of Kronecker products used here, may be found in Marcus and Minc [4].

Theorem 1 may be used to compute the probability density $\{v_k\}$ recursively for given representations $(\underline{\alpha},T)$ and $(\underline{\beta},S)$. It also has the following consequence of interest to the Galton-Watson process.

Theorem 2. Consider a Galton-Watson process in which the number of offspring of an individual has the probability density $\{p_k\}$ of phase type with representation $(\underline{\alpha},T)$, with generating function $P(z)$, given by formula (10). The probability generating function $P_n(z)$ of the number of descendants in the n-th generation of a single progenitor is then given by the n-fold functional iterate of $P(z)$. It follows that the corresponding probability density $\{p_k(n)\}$ is of phase type and has the representation $[\underline{\alpha}(n),T(n)]$, where the order of the matrix $T(n)$ is m^n. The vector $\underline{\alpha}(n)$ and the matrix $T(n)$ are recursively defined by

$$\begin{aligned}
&T(1) = T, \qquad \underline{\alpha}(1) = \underline{\alpha},\\
&T^o(1) = \underline{T}^o, \qquad \alpha^o(1) = \alpha_{m+1},\\
&T(n+1) = T(n) \otimes I_m + T^o(n)A(n) \otimes [I_m - \alpha^o(n)T]^{-1}T,\\
&\underline{T}^o(n+1) = \underline{T}^o(n) \otimes [I_m - \alpha^o(n)T]^{-1}\underline{T}^o,\\
&\underline{\alpha}(n+1) = \underline{\alpha}(n) \otimes \underline{\alpha}[I_m - \alpha^o(n)T]^{-1},\\
&\alpha^o(n+1) = \alpha_{m+1} + \alpha^o(n)\underline{\alpha}[I_m - \alpha^o(n)T]^{-1}\underline{T}^o,
\end{aligned} \tag{20}$$

for $n \geq 1$. $A(n)$ is a diagonal matrix of order m^n with the components of $\underline{\alpha}(n)$ on the diagonal.

Proof. Immediate by repeated application of Theorem 1.

Theorem 2 generalizes the classical result [2], that in a Galton-Watson process in which the density $\{p_k\}$ is geometric, or more generally where $P(z) = p_0 + (1-p_0)pz(1-qz)^{-1}$, all the successive generation sizes have a *geometric* distribution. If $m \geq 2$, the algorithmic utility of the recursive formulas (20) is limited due to the rapid growth in the order of the matrices appearing in the representation. A feasible algorithm to compute the densities of the successive generation sizes is given below. The last of the formulas (20) is computationally useful, since $\alpha^o(n)$ is the probability that the population is extinct by the n-th generation. The sequence $\{\alpha^o(n), n \geq 1\}$ is the probability distribution sequence of the time till extinction.

The following result is useful in certain computations related to the M/G/1 queue. Let X be the length of a random time interval with probability distribution F(·) with Laplace-Stieltjes transform f(s). Consider a Poisson process of rate λ, independent of X, and let N be the number of arrivals in the Poisson process in [0,X].

Theorem 3. If the random variable X has a (continuous parameter) distribution of phase type on $[0,\infty)$, with representation $(\underline{\alpha},T)$, then the random variable N is of phase type on the nonnegative integers with the representation $(\underline{\beta},S)$, given by

$$S = \lambda(\lambda I - T)^{-1}, \quad \underline{\beta} = \underline{\alpha}\lambda(\lambda I - T)^{-1}, \tag{21}$$

and correspondingly

$$\underline{S}^o = (\lambda I - T)^{-1}\underline{T}^o, \quad \beta_{m+1} = \alpha_{m+1} + \underline{\alpha}(\lambda I - T)^{-1}\underline{T}^o. \tag{22}$$

Proof. The probability generating function K(z) of N is given by

$$K(z) = f(\lambda - \lambda z) = \alpha_{m+1} + \underline{\alpha}(\lambda I - \lambda z I - T)^{-1}\underline{T}^o, \tag{23}$$

for $|z| \leq 1$. The function K(z) may be rewritten as

$$\begin{aligned} K(z) = \alpha_{m+1} &+ \underline{\alpha}(\lambda I - T)^{-1}\underline{T}^o \\ &+ z\underline{\alpha}\lambda(\lambda I - T)^{-1}[I - \lambda z(\lambda I-T)^{-1}]^{-1}(\lambda I-T)^{-1}\underline{T}^o. \end{aligned} \tag{24}$$

It suffices to show that formula (24) is a valid representation of the generating function of a density of phase type. Since T is a stable matrix, whose inverse exists, the matrix $\lambda I - T$ is nonsingular. In order to show that $\lambda(\lambda I - T)^{-1}$ is substochastic, consider the probability A_{ij} that at the first event in the Poisson process, the Markov chain of phases is in the state $j \neq m + 1$, given that at time 0 it was in the state $i \neq m + 1$. The matrix $A = \{A_{ij}\}$ is then clearly substochastic. It is explicitly given by

$$A = \int_0^\infty \lambda e^{-\lambda u} e^{Tu} du = \lambda(\lambda I - T)^{-1}. \tag{25}$$

Furthermore

$$\lambda(\lambda I - T)^{-1}\underline{e} + (\lambda I - T)^{-1}\underline{T}^o = (\lambda I - T)^{-1}(\lambda\underline{e} + \underline{T}^o) = \underline{e}, \tag{26}$$

since $\underline{T}^o = -T\underline{e}$. Finally the matrix $I - \lambda(I - T)^{-1}$ is nonsingular, since

$$\begin{aligned} I - \lambda(\lambda I - T)^{-1} &= (\lambda I - T)^{-1}(\lambda I - T - \lambda I) \\ &= -(\lambda I - T)^{-1}T, \end{aligned} \tag{27}$$

and T is nonsingular.

The probability density $\{h_k\}$ of N is given explicitly by

$$\begin{aligned} h_0 &= \alpha_{m+1} + \underline{\alpha}(\lambda I - T)^{-1}\underline{T}^o, \\ h_k &= \lambda^k\underline{\alpha}(\lambda I - T)^{-k-1}\underline{T}^o, \qquad \text{for } k \geq 1. \end{aligned} \tag{28}$$

In the context of the $M^X/G/1$ queue with group arrivals, we consider the case where at each event in the Poisson process, a random group of customers arrives in the queue. If the probability density of the group sizes is also of phase type, it follows from Theorem 1 that the corresponding random variable N again has a density of phase type, since its generating function is given by $K[\phi(z)]$, where $\phi(z)$ denotes the probability generating function of the group sizes. The representation of the density of N can be readily constructed from that of $K(z)$ and $\phi(z)$ by application of Theorem 1.

Formula (28) permits a particularly simple recursive computation of the density $\{h_k\}$ and therefore of the stationary queue length distribution in an M/G/1 queue in which the service time distribution is of phase type. In [8] it is further shown that the stationary distribution of the FIFO waiting time is itself of phase type and may be computed by solving a linear system of differential equations with constant coefficients.

2. THE SUCCESSIVE GENERATION SIZES IN A GALTON-WATSON PROCESS

Except for particularly simple densities of the number of offspring per individual, the probability densities of the successive generation sizes are not easily computed. If $\{p_k(n)\}$ is the probability density of the size of the n-th generation, then we have

$$p_k(1) = p_k,$$
$$p_k(n+1) = \sum_{\nu=0}^{\infty} p_\nu(n) p_k^{(\nu)}, \quad \text{for } k \geq 0. \tag{29}$$

Even if the original density $\{p_k\}$ is concentrated on the integers $0,\ldots,M$, the density $\{p_k(n)\}$ will be concentrated on the integers $0,\ldots,M^n$. For larger values of n, this will require a truncation of the density $\{p_k(n)\}$ with a resulting error which will be propagated in the computation of $\{p_k(n+1)\}$. This is a fortiori the case when the density $\{p_k\}$ does not have a bounded support. The rapid growth of the number of points in the support of $\{p_k(n)\}$ also appears to exclude the use of the fast Fourier transform as a feasible computational technique.

If the density $\{p_k\}$ is of phase type, it is possible to construct an algorithm which, at least for subcritical processes, is highly efficient and accurate. The main ingredient of the algorithm is a recursive procedure for the computation of the density of finite mixtures of the general type.

$$\underline{r} = a_0 + a_1\underline{p} + a_2\underline{p}^{(2)} + \ldots + a_N\underline{p}^{(N)}, \tag{30}$$

where $\underline{p}^{(k)}$ denotes the k-fold convolution of the probability density $\underline{p}$, and a_j, $0 \leq j \leq N$, are probabilities whose sum is one.

The probability density $\underline{r}$ is itself of phase type and we now construct its representation. Let the density $\underline{p}$ have the representation $(\underline{\alpha},T)$, where T is a matrix of order m. The density $\underline{r}$ will be represented as the probability density of the time till absorption in a Markov chain with Nm + 1 states. For notational convenience we label the absorbing state 0 and the transient states 1,...,Nm. It will be further convenient to think in terms of an urn model, in which at time n = 0, a random number J with $P\{J = j\} = a_j$, $0 \leq j \leq N$, of particles are placed. At time 0, we also "start" a Markov chain with m + 1 states and transition probability matrix B, given by formula (3). The initial state is chosen with probabilities $\alpha_1,\ldots,\alpha_m$, α_{m+1}. If the state m + 1 is drawn, a particle is removed from the urn and a new independent multinomial trial is performed. This is continued until either the urn is empty or until an initial state other than m + 1 is selected. This procedure determines the content of the urn at time n = 0+. If at time n = 0+, the urn is not empty, we consider the Markov chain B at successive time points n = 1,2,... As long as states other than m + 1 are visited, no particles are removed from the urn. Whenever the state m + 1 is reached, a particle is removed from the urn and an "instantaneous" sequence of multinomial trials is performed with probabilities $\alpha_1,\ldots,\alpha_m$, α_{m+1}. Whenever the state m + 1 appears a particle is removed from the urn. The absorbing chain is restarted in this manner until the urn becomes empty. In order to construct the representation for $\underline{r}$, we consider the number J_n of particles in the urn at time n+, and the state I_n of the Markov chain B at time n+.

Clearly $0 \leq J_n \leq N$, and $1 \leq I_n \leq m$. If $J_n = 0$, I_n is not defined. We shall say that the Markov chain, associated with $\underline{r}$

(the large chain) is in the state $(i - 1)m + j$, $1 \leq i \leq N$, $1 \leq j \leq m$, at time n, if and only if $J_n = i$, $I_n = j$. It is clear that $\underline{r}$ is the probability density of the time until the urn becomes empty.

If we denote the representation of $\underline{r}$ by $(\underline{\gamma}, T^*)$, then $\underline{\gamma}$ is given by

$$\gamma_{(i-1)m+j} = \gamma_{ij} = \sum_{\nu=i}^{N} a_\nu \alpha_{m+1}^{\nu-i} \alpha_j, \text{ for } 1 \leq i \leq N,\ 1 \leq j \leq m$$

$$\gamma_0 = \sum_{\nu=0}^{N} a_\nu \alpha_{m+1}^{\nu}. \tag{31}$$

The matrix T*, which is of order mN, is of a block lower triangular form and may be written as

$$T^* = \begin{vmatrix} T & 0 & 0 & \cdots & 0 \\ T^0 A & T & 0 & \cdots & 0 \\ \alpha_{m+1} T^o A & T^o A & T & \cdots & 0 \\ \alpha_{m+1}^2 T^o A & \alpha_{m+1} T^o A & T^o A & \cdots & 0 \\ \vdots & \vdots & \vdots & & \vdots \\ \alpha_{m+1}^{N-2} T^o A & \alpha_{m+1}^{N-3} T^o A & \alpha_{m+1}^{N-4} T^o A & \cdots & T \end{vmatrix} \tag{32}$$

The corresponding column vector $\underline{T}^{*o}$ is given by $[\underline{T}^o, \alpha_{m+1}\underline{T}^o, \ldots \alpha_{m+1}^{N-1}\underline{T}^o]$.

The computation of the distribution of the density $\underline{r}$ is now equivalent to that of the probability distribution of the time till absorption in the $(mN + 1)$-state Markov chain

$$P^* = \begin{pmatrix} 1 & \underline{0} \\ \underline{T}^{*o} & T^* \end{pmatrix}, \tag{33}$$

with the initial probability vector $(\gamma_0, \underline{\gamma})$. Consider a sequence of vectors $\underline{v}(n) = \{v_0(n), \underline{v}^{(1)}(n), \ldots, \underline{v}^{(N)}(n)\}$, of dimension $mN + 1$, defined by

$$v_0(0) = \gamma_0, \quad \{\underline{v}^{(1)}(0),\ldots,\underline{v}^{(N)}(0)\} = \underline{\gamma},$$
$$v_0(n+1) = v_0(n) + \sum_{\nu=1}^{N} \alpha_{m+1}^{\nu-1}\underline{v}^{(\nu)}(n)\underline{T}^{o}, \tag{34}$$
$$\underline{v}^{(i)}(n+1) = \underline{v}^{(i)}(n)T + \sum_{\nu=i+1}^{N} \alpha_{m+1}^{\nu-i-1}\underline{\alpha}[\underline{v}^{(\nu)}(n)\underline{T}^{o}],$$

for $1 \leq i \leq N$, and $n \geq 0$, then it is clear that the sequence $\{v_0(n),\ n \geq 0\}$ is the probability distribution corresponding to the required density. The recurrence relations (34) may be easily programmed for numerical computation. Since $v_0(n)$ tends to one, the sum of the other components of the vector $\underline{v}(n)$ tends to zero. In particular we have $\underline{v}^{(N)}(n+1) = \underline{v}^{(N)}(n)T = a_N\underline{\alpha}T^{n+1}$. For larger values of n, the recurrence relations (34) involve many computational steps which contribute only negligable amounts to the terms $v_0(n)$. Many of these steps can be eliminated at the expense of a small error but with a significant reduction in computation time. This aspect is discussed below, but for use in the sequel, we first compute the inverse of the matrix $I - T^*$.

Theorem 4. The inverse J of the matrix $I - T^*$ is a block lower triangular matrix whose entries are given by

$$J(i,i) = (I - T)^{-1}, \quad \text{for } 1 \leq i \leq N,$$
$$J(i,j) = D(I - T)^{-1}, \quad \text{for } i > j > 1. \tag{35}$$

The matrix $D = (I - T)^{-1}T^{o}A$, has identical rows, all equal to $\underline{\alpha}$.

Proof. Since $(I - T)^{-1}\underline{T}^{o} = \underline{e}$, it is clear that D has the stated property. The diagonal blocks of $(I - T^*)J$ are all equal to I. Computing the offdiagonal blocks we obtain for $i > j > 1$, that

$$[(I-T^*)J](i,j) = -\alpha_{m+1}^{i-j-1}T^{o}A(I-T)^{-1}$$
$$- \sum_{\nu=0}^{i-j-2} \alpha_{m+1}^{i-j-\nu}T^{o}AD(I-T)^{-1} + (I-T)D(I-T)^{-1}. \tag{36}$$

Since $T^{o}AD = (1 - \alpha_{m+1})T^{o}A$, it readily follows that all the off-diagonal blocks are zero.

2.1. *Adaptive Trimming*

In the recursive computation of the vectors $\underline{v}(n)$ by means of the recurrence relations (34), the sums of the upper components of the nonnegative vectors $\underline{v}^{(i)}(n)$ for n large tend rapidly to zero. It is to our advantage to reduce the value of N for appropriate values of n, thereby saving a significant number of arithmetic operations in the computation of $\underline{r}$.

In a first method, we determine for each $n > 0$, the index $N_1(n) = \max_i\{i: \underline{v}^{(i)}(n)\cdot\underline{e} \geq \varepsilon\}$ and implement the recurrence relations for $n + 1$ with N replaced by $N_1(n)$. If we denote the reduced matrix following the first trimming by $T^*(N_1)$, then the amount of probability neglected in the tail of the distribution after the first trimming is given by

$$\underline{\gamma}T^{*n-1}(N)\underline{e} - \underline{\gamma}(N_1)T^{*n-1}(N_1)\underline{e}(N_1)$$
$$= \sum_{i=N_1+1}^{N} \underline{v}^{(i)}(n)\underline{e} < (N - N_1)\varepsilon. \tag{37}$$

Continuing this procedure until $N_1(n)$ reaches zero, we obtain the computed sequence $\{\hat{v}_0(n), n \geq 0\}$, which is nondecreasing and satisfies $\hat{v}_0(n) \leq v_0(n)$ for all $n \geq 0$ and $1 - \hat{v}_0(\infty) < \varepsilon N$. An upper bound on the error in each of the density terms is $2\varepsilon N$, but this bound is very conservative.

The first trimming method is undesirable in repeated applications of the algorithm, as needed in computations for the Galton-Watson process, since the computed density corresponding to $\{\hat{v}_0(n)\}$ plays the role of the mixing density $\underline{a}$ for the next generation. The number of terms computed varies from one generation to the next and the accumulated error grows in a generally unpredictable manner. Moreover the computed distributions are defective.

A more conservative trimming procedure is the following: Let $N_2(0) = N$, and determine for each $n \geq 1$, the index $N_2(n)$ by

$$N_2(n) = \max_{1 \leq i \leq N_2(n-1)} \{i: \sum_{j \geq i} (j-i+1)\hat{\underline{v}}^{(j)}(n)\cdot\underline{e} \geq \varepsilon\}, \tag{38}$$

where $\hat{\underline{v}}^{(i)}(n)$ is the computed value of $\underline{v}^{(i)}(n)$. If the set in braces is empty, set $\hat{v}_0(n+1) = 1$, and stop. If $N_2(n) = N_2(n-1) \geq 1$, implement the recurrence relations (34) with $N = N_2(n)$. If $N_2(n-1) > N_2(n) \geq 1$, replace the vector $\hat{\underline{v}}(n)$ by the vector

$$\hat{\underline{v}}^{(1)}(n),\ldots,\hat{\underline{v}}^{(N_2(n)-1)}(n), \quad \sum_{i=N_2(n)}^{N_2(n-1)} \hat{v}^{(i)}(n),\underline{0},\ldots,\underline{0} \tag{39}$$

and implement (34) with $N = N_2(n)$. This method has the advantage that no probability is "lost" in the recursive computation in the sense that for all $n \geq 0$, we have that

$$\hat{v}_0(n) + \sum_{i=1}^{N} \hat{\underline{v}}^{(i)}(n)\underline{e} = 1. \tag{40}$$

The computed values $\hat{v}_0(n)$ now satisfy $\hat{v}_0(n) \geq v_0(n)$, for all $n \geq 0$, and the computed sequence $\{\hat{v}_0(n)\}$ is a probability distribution concentrating on a finite number of nonnegative integers.

An appropriate measure of the truncation error is the quantity

$$\Delta(\varepsilon) = \sum_{n=0}^{\infty} [\hat{v}_0(n) - v_0(n)] = \sum_{n=0}^{\infty} [1 - v_0(n)] - \sum_{n=0}^{\infty} [1 - \hat{v}_0(n)]. \tag{41}$$

We see that $\Delta(\varepsilon)$ is the difference between the exact mean of the desired distribution $\{v_0(n)\}$ and the mean of the computed distribution $\{\hat{v}_0(n)\}$.

We shall now obtain an estimate of the quantity $\Delta(\varepsilon)$, and to this end we first make a number of preliminary observations.

Let the first trimming occur after the computation of $\underline{v}(n)$ and let it reduce N to N'. The modified vector defined in (42) may then be written as $\underline{v}(n)Z_{N'}$, where the matrix $Z_{N'}$ is defined as an $N \times N$ matrix of $m \times m$ blocks, with

$$\begin{aligned} Z_{N'}(i,i) &= I_m, && \text{for } 1 \leq i \leq N' \\ Z_{N'}(i,N') &= I_m, && \text{for } N' \leq i \leq N \\ Z_{N'}(i,j) &= 0, && \text{for all other pairs.} \end{aligned} \tag{42}$$

The mean M of the density $\underline{r}$ is given by

$$\begin{aligned} M &= \underline{\gamma}[I - T^*(N)]^{-1}\underline{e} \\ &= \underline{\gamma} \cdot \sum_{k=0}^{n-1} T^{*k}(N)\underline{e} + \underline{\gamma}T^{*n}(N)[I - T^*(N)]^{-1}\underline{e}. \end{aligned} \tag{43}$$

Following the first trimming, the mean of the computed distribution is reduced to

$$M_1 = \underline{\gamma} \sum_{k=0}^{n-1} T^{*k}(N)\underline{e} + \underline{\gamma}T^{*n}(N)Z_{N'}[I - T^*(N)]^{-1}\underline{e}. \tag{44}$$

Using the explicit form of the inverse, obtained in Theorem 4, we obtain that

$$[I - T^*(N)]^{-1}\underline{e} = \begin{pmatrix} (I-T)^{-1}\underline{e} \\ (I-T)^{-1}\underline{e} + \mu\underline{e} \\ (I-T)^{-1}\underline{e} + 2\mu\underline{e} \\ \vdots \\ (I-T)^{-1}\underline{e} + (N-1)\mu\underline{e} \end{pmatrix}, \tag{45}$$

so that

$$Z_{N'}[I - T^*(N)]^{-1}\underline{e} = \begin{pmatrix} (I-T)^{-1}\underline{e} \\ (I-T)^{-1}\underline{e} + \mu\underline{e} \\ \vdots \\ (I-T)^{-1}\underline{e} + (N' - 1)\mu\underline{e} \\ \vdots \\ (I-T)^{-1}\underline{e} + (N' - 1)\mu\underline{e} \end{pmatrix} \tag{46}$$

It follows that

$$M - M_1 = \mu \sum_{i=1}^{N-N'} i\underline{v}^{(N'+i)}\underline{e} < \mu\varepsilon. \tag{47}$$

Since the computation after the first trimming is similar in nature to the original one, we see that the mean of the computed distribution differs from the mean M by at most $N\mu\varepsilon$, so that

$$\Delta(\varepsilon) < N\mu\varepsilon. \tag{48}$$

Remark. It is of course possible to implement the original recurrence relations (34) up to the smallest index n^* for which

$$\sum_{n=0}^{n^*} [1 - v_0(n)] > M - \varepsilon, \tag{49}$$

thereby guaranteeing that the mean of the computed distribution differs from the exact one by at most ε. The advantage of the adaptive trimming procedure lies in the progressive reduction of the number of operations involved in the recurrence relations (34), which is particularly significant for *stable* Galton-Watson processes.

2.2. *The Successive Generation Sizes*

The probability densities $\{p_k(n)\}$ of the successive generation sizes of a Galton-Watson process, in which the density $\{p_k\}$ is of phase type, may be computed by repeated applications of the algorithm developed above for the mixtures defined by formula

(30). We note that the recurrence relations (29) are valid for a single progenitor. If there are ν progenitors with probability a_ν, $0 \leq \nu \leq N$, then the first equation in (29) should be replaced by

$$p_0(1) = \sum_{\nu=0}^{N} a_\nu p_0^\nu,$$
$$p_k(1) = \sum_{\nu=1}^{N} a_\nu p_k^{(\nu)}, \qquad \text{for } k \geq 1. \tag{50}$$

The computation of $\{p_k(1)\}$ is clearly of the type defined by formula (30). It results in a computed density $\{\hat{p}_k(1),\ 0 \leq k \leq N_1\}$, which plays the role of the density $\{a_\nu\}$ in the computation of $\{p_k(2)\}$ and so on. By using the second adaptive trimming procedure, discussed above, we may use *the computed means* $M_1(n)$, $n \geq 0$, of the successive generation sizes to keep track of the accumulated truncation and trimming errors. The means $M(n)$ of the exact distributions are of course given by

$$M(n) = \sum_{\nu=1}^{N} \nu a_\nu \mu^n, \qquad \text{for } n \geq 0. \tag{51}$$

For Galton-Watson processes for which μ is significantly less than one and the maximum initial population size N not too large, this method permits us to study the successive generation sizes until the extinction probability $p_0(n)$ becomes close to one. Computation times are generally small, on the order of a few seconds per generation. If μ is close to one, and a fortiori when μ is greater than one, the support of the successive computed densities increases with n and the computation time per generation increases quite rapidly.

We also note that this computational method needs to be only trivially modified to handle cases where the probability density of the number of offspring depends on the index of each generation. Immigration or removals from the population can also be studied by routine modifications of the algorithm.

The mean μ alone does not provide much information on the size of possible large excursions of the Galton-Watson process before extinction. It is possible in many cases however, to compute the distribution of the maximum generation size before extinction. The appropriate algorithm is discussed in Section 4.

The matrix T is usually very sparse and a major reduction in the computation time can be achieved by writing special purpose routines to compute the products $\underline{v}^{(i)}(n)T$ in the last formula (34).

3. A GALTON-WATSON PROCESS EMBEDDED IN THE $M^X/G/1$ QUEUE

Consider an $M^X/G/1$ queue with group arrivals and let the probability generating function of the density $\{a_k\}$ of the group sizes be $\phi(z)$, with (0) = 0. The arrival rate (of groups) is λ and the service time distribution is denoted by $H(\cdot)$ with Laplace-Stieltjes transform $h(s)$. If the mean service time is μ and the mean group size is η, then it is well-known that the queue is stable if and only if $\lambda\eta\mu \leq 1$. Stationary distributions of the relevant queue features exist if and only if $\lambda\eta\mu < 1$.

Let $t = 0$, be the beginning of a service and let the queue length $\xi(0)$ at $t = 0$, be equal to i_0. Let T_1 be the time when all customers present at $T_0 = 0$, have been served under the FIFO discipline and let $\xi(T_1)$ denote the number of arrivals during the interval (T_0,T_1). Similarly T_2 is the time when all $\xi(T_1)$ customers present at time T_1+ have been served and $\xi(T_2)$ denotes the number of arrivals in (T_1,T_2). This construction is repeated to yield a bivariate sequence of random variables $\{(T_n,\xi(T_n), n \geq 0\}$. We shall agree that if $\xi(T_n) = 0$, then T_{n+1} is the time when the group of customers, who arrive during the idle period starting at T_n, have completed service. The marginal sequence $\{\xi(T_n), n \geq 0\}$ is known to be a Markov chain on the nonnegative integers, [3,7]. Its transition probability matrix U is given by

$$U = \begin{vmatrix} a'_0 & a'_1 & a'_2 & a'_3 & \cdots \\ p_0 & p_1 & p_2 & p_3 & \cdots \\ p_0^{(2)} & p_1^{(2)} & p_2^{(2)} & p_3^{(2)} & \cdots \\ p_0^{(3)} & p_1^{(3)} & p_2^{(3)} & p_3^{(3)} & \cdots \\ \vdots & \vdots & \vdots & \vdots & \end{vmatrix}, \tag{52}$$

where $\{p_k\}$ is the probability density with generating function $P(z) = h[\lambda-\lambda\phi(z)]$, $\{p_k^{(\nu)}\}$ is its ν-fold convolution, and the density $\{a'_k\}$ is defined by

$$a'_k = \sum_{j=1}^{\infty} a_j p_k^{(j)}, \qquad \text{for } k \geq 0. \tag{53}$$

The probability generating function $A(z)$ of $\{a'_k\}$ is clearly given by

$$A(z) = \phi\{h[\lambda-\lambda\phi(z)]\}, \qquad \text{for } |z| \leq 1. \tag{54}$$

We note that if $H(\cdot)$ is a (continuous) distribution of phase type and $\{a_k\}$ a (discrete) density of phase type, then by Theorem 3, $\{p_k\}$ is of phase type, and by Theorem 1, $\{a'_k\}$ is of phase type. Representations for $\{p_k\}$ and $\{a'_k\}$ may easily be constructed from those of $H(\cdot)$ and $\{a_k\}$.

Assuming henceforth that $\lambda\eta\mu < 1$, we proceed to discuss the stationary density $\{\pi_k\}$ of the recurrent Markov chain U. The quantities π_k, $k \geq 0$, satisfy the system of equations

$$\begin{aligned} &\pi_0 a' + \sum_{j=1}^{\infty} \pi_j p_\nu^{(j)} = \pi_\nu, \qquad \text{for } \nu \geq 0 \\ &\sum_{j=0}^{\infty} \pi_j = 1. \end{aligned} \tag{55}$$

Denoting the generating function of $\{\pi_j\}$ by $\pi(z)$, we obtain

$$\pi(z) = \pi[P(z)] - \pi_0[1 - A(z)], \qquad \text{for } |z| \leq 1. \tag{56}$$

Theorem 5. The probability generating function $\pi(z)$ is given by

$$\pi(z) = 1 - \{1 + \sum_{j=0}^{\infty} [1-A[P_j(0)]]\}^{-1} \sum_{\nu=0}^{\infty} [1-A[P_\nu(z)]], \quad \text{for } |z| \leq 1, \tag{57}$$

where $P_\nu(z)$ is the ν-th functional iterate of $P(z)$. $P_0(z) = z$.

Proof. Replacing z by $P_n(z)$ in (56), we obtain

$$\pi[P_n(z)] = \pi[P_{n+1}(z)] - \pi_0[1-A[P_n(z)]], \quad \text{for } n \geq 0, \tag{58}$$

and hence

$$\pi(z) = \pi[P_{n+1}(z)] - \pi_0 \sum_{j=0}^{n} [1-A[P_j(z)]]. \tag{59}$$

Since $P'(1) = \lambda\eta\mu < 1$, we know that $P_n(z) \to 1$, for all $0 \leq z \leq 1$, as n tends to infinity. The series of analytic functions

$$\sum_{j=0}^{\infty} [1-A[P_j(z)]]$$

converges uniformly for all $0 \leq z \leq 1$. This follows from the Lebesgue dominated convergence theorem, since

$$1-A[P_j(z)] \leq 1-A[P_j(0)] \leq \theta_j, \tag{60}$$

where

$$\theta_j = [\frac{d}{dz} A[P_j(z)]]_{z=1} = \eta(\lambda\eta\mu)^{j+1}.$$

The second inequality in (60) is obtained by noting that the graph of the convex increasing function $A[P_j(z)]$ lies for every in $0 \leq z < 1$, above its tangent at $z = 1$.

Passing to the limit in (59), we obtain

$$\pi(z) = 1 - \pi_0 \sum_{j=0}^{\infty} [1-A[P_j(z)]], \tag{61}$$

for $0 \leq z \leq 1$. By analytic continuation, the same formula is valid for $|z| \leq 1$. Setting $z = 0$, in formula (61) we obtain

$$\pi_0 = \{1 + \sum_{j=0}^{\infty} [1-A[P_j(0)]]\}^{-1}. \tag{62}$$

Remark. We note that

$$\sum_{j=0}^{\infty} [1-A[P_j(0)]] = d, \tag{63}$$

is the mean number of generations till extinction in a Galton-Watson process with offspring density $\{p_k\}$ and initial population size density $\{a_k'\}$. Applying the second inequality in (60) we obtain

$$d \leq \frac{\lambda\eta^2\mu}{1-\lambda\eta\mu}, \tag{64}$$

and hence

$$\pi_0 \geq \frac{1-\lambda\eta\nu}{1-\lambda\eta\mu(1-\eta)}. \tag{65}$$

By differentiating k times in (61), we obtain the explicit formula

$$\pi_k = \pi_0 \sum_{j=0}^{\infty} P(j,k), \qquad \text{for } k \geq 1, \tag{66}$$

where $P(j,k)$ is the probability that there are k individuals in the j-th generation of a Galton-Watson process with offspring density $\{p_k\}$ and initial population density $\{a_k'\}$. The initial population is counted as generation 0.

3.1. *Computational Aspects*

The density $\{\pi_k\}$ may be accurately computed for queues for which the underlying distributions are of phase type, by means of the recursive algorithm developed in Section 2; this at least if $\lambda\eta\mu$ is not too close to one. The quantities $P_j(0)$ can be efficiently computed by successive substitutions in the probability generating function, but each step involves two matrix inversions. The value of π_0 is computed in terms of the $P_j(0)$. The term-wise sums of the densities $\{P(j,k), k \geq 1\}$ over the index j are formed and serve in the computation of π_k, for $k \geq 1$. It is advisable to compute π_0 separately and to a high accuracy. If the value of π_0 is essentially correct, we may use the normalizing condition $\Sigma\pi_k = 1$, to determine the number of generations needed in (66) to obtain the probabilities π_k, $k \geq 1$, to a sufficient degree of accuracy. The computation of the density $\{\pi_k\}$ is of interest in the numerical investigation of the priority rules discussed by Nair and Neuts [5,6].

4. THE MAXIMUM GENERATION SIZE BEFORE EXTINCTION

The random variable $Y = \max\{X_n, n \geq 0\}$ of the successive generation sizes before extinction in a Galton-Watson process has been discussed by J. Bishir [1] and E. Seneta [9].

For each $k \geq 1$, the system of linear equations

$$y_i^{(k)} = \sum_{\nu=1}^{k} p_\nu^{(i)} y_\nu^{(k)} + p_0^i, \quad \text{for } 1 \leq i \leq k, \tag{67}$$

has a unique solution $[y_1^{(k)}, \ldots, y_k^{(k)}]$ and

$$P\{Y \leq k\} = \sum_{i=1}^{k} P\{X_0 = i\} y_i^{(k)}, \quad \text{for } k \geq 1. \tag{68}$$

For a subcritical or critical Galton-Watson process the distribution of Y is honest, but for a supercritical process we have

$$P\{Y < \infty\} = \sum_{i=1}^{\infty} P\{X_0 = i\} \rho^i, \tag{69}$$

where ρ is the probability of extinction for the line of a single progenitor.

Bishor's paper does not enter into the construction of an efficient algorithm for the computation of the distribution of Y. The examples of highly subcritical or highly supercritical cases, presented in [1], are somewhat misleading in assessing the computational effort involved. We examined the following two methods for a large number of examples:

4.1. *The Gauss-Seidel Method*

For each k, the system of linear equations (67) satisfies sufficient conditions for the convergence of the Gauss-Seidel iterative method. It is easy to show that the quantities $y_i^{(k)}$ satisfy $y_1^{(k)} \geq y_2^{(k)} \geq \ldots \geq y_k^{(k)}$, since $y_i^{(k)}$ is the probability that a Galton-Watson process with i progenitors becomes extinct without exceeding the population size k. After solving the system of equations for k, it is convenient to use the (k+1)-tuple $y_1^{(k)},\ldots,y_k^{(k)}, y_k^{(k)}$, as a starting solution for the computation of the quantities $y_1^{(k+1)},\ldots,y_{k+1}^{(k+1)}$.

For Galton-Watson processes which are close to critical, and in general when systems in excess of k = 75, need to be solved, the computation time for the Gauss-Seidel method becomes substantial and exceeds one minute of central processing time on a CDC 6500 computer.

4.2. *The Gauss Elimination Method*

Writing the system (67) as

$$\sum_{\nu=1}^{k} (\delta_{i\nu} - p_\nu^{(i)})y_\nu^{(k)} = p_0^i, \qquad 1 \leq i \leq k, \tag{70}$$

assume that the system has been reduced to upper triangular form by elementary row operations represented by the lower triangular matrix K_k. The resulting system is written in the form

$$H_k \underline{y}^{(k)} = \underline{c}^{(k)}, \tag{71}$$

where H_k is upper triangular with $H_{k,11} = 1$. The system (71) is readily solved and $P\{Y \leq k\}$ is computed.

The appealing feature of this method is the easy computation of the matrix H_{k+1} and the vector $\underline{c}_{k+1}$. This is described in the following algorithmic steps to go from k to k+1:

Step 1: Compute p_{k+1}.

Step 2: Compute $p_{k+1}^{(j)}$, for $j = 1,\ldots,k$.

Step 3: Compute the terms p_0^{k+1} and $p_i^{(k+1)}$, for $1 \leq i \leq k+1$.

Step 4: Left-multiply the vector computed in Step 2 by the matrix K_k, to obtain the first k entries in the (k+1)-st column of H_{k+1}.

Step 5: Perform Gauss elimination on the row computed in Step 3, to obtain the (k+1)-st row of H_{k+1} and the (k+1)-st entry of $\underline{c}_{k+1}$.

Step 6: Compute $y_i^{(k+1)}$, for $1 \leq i \leq k+1$.

Step 7: Compute $P\{Y \leq k+1\}$. If $P\{Y \leq k+1\}$ is sufficiently close to the probability of eventual extinction, stop. If not, set k equal to k + 1 and go to Step 1.

This method is much faster than the Gauss-Seidel method, but may be sensitive to the accuracy problems usually associated with Gauss elimination. The strong diagonal dominance of the coefficient matrix in the system (70) suggests that these problems will be minor. Both methods were compared in single precision on the CDC 6500, which is a computer with large word length. Even in examples where k ran up to one hundred, all computed probabilities agreed to at least four decimal places, but on computers with a shorter word length it is probably advisable to perform the latter method in double precision.

For the computation of $P\{Y \leq k\}$, $k \geq 1$, there is no particular advantage in assuming that $\{p_j\}$ is of phase type, except for the easy computation of the terms of the density. A minor drawback of the Gauss elimination lies in the substantial storage requirements. If we allow values of k up to one hundred, two storage arrays of size 10000 are required, one to store the quantities $\{p_i^{(j)}\}$ and the second one to store the entries of the matrices K_k and H_k.

5. THE PROBABILITY OF EVENTUAL ABSORPTION

It is well-known that the probability of extinction of the lineage of a single progenitor is given by the smallest positive root ρ of the equation

$$z = \alpha_{m+1} + z\underline{\alpha}(I - zT)^{-1}\underline{T}^{o}, \tag{72}$$

and that $\rho = 1$, if and only if $\mu = \underline{\alpha}(I - T)^{-1}\underline{e} \leq 1$. For $\mu > 1$, we may compute ρ by successive substitutions or more efficiently by Newton's method. Since the derivative of the right hand side is given by

$$P'(z) = \underline{\alpha}(I - zT)^{-2}\underline{T}^{o}, \tag{73}$$

the successive Newton approximations are given by

$$\begin{aligned} z_{\nu+1} &= [1-\underline{\alpha}(I-z_{\nu}T)^{-2}\underline{T}^{o}]^{-1}[\alpha_{m+1}+z_{\nu}\underline{\alpha}(I-z_{\nu}T)^{-1}\underline{T}^{o}-z_{\nu}\underline{\alpha}(I-z_{\nu}T)^{-2}\underline{T}^{o}] \\ &= [1-\underline{\alpha}(I-z_{\nu}T)^{-2}\underline{T}^{o}]^{-1}[\alpha_{m+1}-z_{\nu}^{2}\underline{\alpha}(I-z_{\nu}T)^{-2}T\,\underline{T}^{o}] \end{aligned} \tag{74}$$

Since the function $P(z)$ is convex increasing the sequence $\{z_{\nu}\}$, $0 \leq z_0 < 1$, always converges to ρ. Caution is needed when μ is very close to one, since in this case the first factor in (74) becomes very large. Note that $1 - \underline{\alpha}(I-T)^{-2}\underline{T}^{o} = 1 - \mu$. In all other cases Newton's method converges rapidly. Each iteration may be most efficiently computed as follows:

Step 1: Compute $(I - z_{\nu}T)^{-1}$.

Step 2: Compute $\underline{\alpha}(I - z_{\nu}T)^{-1} = \underline{y}$.

Step 3: Evaluate $\underline{u} = \underline{y}(I - z_{\nu}T)^{-1}$.

Step 4: Evaluate $1 - \underline{u}\,\underline{T}^{o}$, and the second factor in (74) and compute $z_{\nu+1}$.

REFERENCES

[1] J. Bishir, Maximum Population Size in a Branching Process, *Biometrics, 18,* pp. 394-403, 1962.

[2] T. E. Harris, "The Theory of Branching Processes," Springer-Verlag, Berlin, 1963.

[3] D. G. Kendall, Some Problems in the Theory of Queues, *J.R.S.S., B, 13,* pp. 151-185, 1951.

[4] M. Marcus and H. Minc, "A Survey of Matrix Theory and Matrix Inequalities," Allyn & Bacon, Boston, 1964.

[5] S. S. Nair and M. F. Neuts, A priority Rule based on the Ranking of the Service Times for the M/G/1 Queue, *Oper. Research, 17,* pp. 466-477, 1969.

[6] S. S. Nair and M. F. Neuts, An Exact Comparison of the Waitingtimes under Three Peiority Rules, *Oper. Research, 19,* 414-423, 1971.

[7] M. F. Neuts, The Queue with Poisson Input and General Service Times, Treated as a Branching Process, *Duke Math. Jour., 36,* pp. 215-232, 1969.

[8] M. F. Neuts, Probability Distributions of Phase Type, Liber Amicorum Prof. Emeritus H. Florin, Dept. of Math., Univ. Louvain, Belgium, 173-206, 1975.

[9] E. Seneta, On the Maxima of Absorbing Markov Chains, *Australian J. of Stat., 9,* pp. 93-102, 1967.

CENTRAL LIMIT ANALOGUES FOR MARKOV POPULATION PROCESSES

Donald R. McNeil[1]

Department of Statistics
The Johns Hopkins University
Baltimore, Maryland

Siegfried Schach[2]

Department of Statistics
The Johns Hopkins University
Baltimore, Maryland

[1]*Research supported by a generous grant from the Ford and Rockefeller Foundations, awarded to the Office of Population Research, Princeton University.* *Present address of author: School of Economics and Financial Management, Macquarie University, North Ryde, 2113 New South Wales, Australia.*

[2]*Research supported by the United States Department of Transportation, Federal Highway Administration, Bureau of Public Roads under Contract DOT-FH-11-7716 awarded to the Department of Statistics, The Johns Hopkins University.*

ISBN 0-12-394680-8

1. SUMMARY

In this paper the main discussion is concerned with obtaining asymptotic results for sequences of birth and death processes which are similar to the central limit theorem for sequences of univariate random variables. The motivation is the need to obtain useful approximations to the distributions of sample paths of processes which arise as models for population growth, but for which Kolmogorov differential equations are intractable.

In the first section, univariate processes are considered, and conditions are given for the weak convergence of

$$Z_N(t) = \{X_N(t) - aN\}/\sqrt{N},$$

where $\{X_N(t), N = 1,2,...\}$ is a sequence of ergodic birth and death processes, to those of an Ornstein-Uhlenbeck process $N \to \infty$. A heuristic method is given which may help explain why this convergence holds, and some examples are given for purposes of illustration.

The second part deals with multivariate processes, and three examples are considered in detail: a model for the growth of the sexes in a biological population, a multivariate Ehrenfest process, and a model for the growth and interreaction of two cities.

The paper concludes with a discussion of various related results. It is shown that in certain special cases it is possible to obtain diffusions other than the Ornstein-Uhlenbeck process as limits. Finally, heavy traffic results are included for congestion situations originally considered in the special case of time-homogenous arrival rates by Kingman. Transient processes such as epidemics are also shown to exhibit a "central limit" behavior.

2. A CENTRAL LIMIT THEOREM FOR UNIVARIATE PROCESSES

Suppose that $X(t)$ is a birth and death process, that is, a continuous time, non-negative integer valued, Markov process, with transition probabilities, $P[X(t+\delta t) - X(t) = j | X(t) = n]$, given by

$$\begin{cases} \lambda(n)\delta t + o(\delta t) & \text{if } j = 1, \\ \mu(n)\delta t + o(\delta t) & \text{if } j = -1 \\ 1 - \{\lambda(n) + \mu(n)\}\delta t + o(\delta t) & \text{if } j = 0 \\ o(\delta t), & \text{otherwise} \end{cases} \tag{2.1}$$

Let us introduce a parameter, N, taking positive integer values, and allow $\lambda(n)$ and $\mu(n)$ to depend on it. Thus we are concerned with a sequence of birth and death processes, $\{X_N(t)\}$, with birth and death rates $\lambda_N(n)$ and $\mu_N(n)$, respectively. We assume that for some $a > 0$, these rates possess the asymptotic expansions

$$\begin{cases} \lambda_N(aN + x\sqrt{N}) = \alpha(a)N + \beta_1(a)x\sqrt{N} + O(1), \\ \mu_N(aN + x\sqrt{N}) = \alpha(a)N + \beta_2(a)x\sqrt{N} + O(1). \end{cases} \tag{2.2}$$

We now consider the transformed process

$$Z_N(t) = \{X_N(t) - aN\}/\sqrt{N}. \tag{2.3}$$

In his 1961 Ph.D. dissertation at Stanford University, C. Stone discussed the questions of convergence of Markov processes, obtaining in particular conditions for the weak convergence of $Z_N(t)$ to a diffusion process. The relevant result, quoted by Iglehart (1965; Theorem 3.2), is that provided the initial distributions converge, $Z_N(t)$ converges weakly as $N \to \infty$ to an Ornstein-Uhlenbeck process $Z(t)$ characterized by the forward Kolmogorov equation

$$\frac{\partial f}{\partial x} = \alpha(a)\frac{\partial^2 f}{\partial x^2} + \{\beta_2(a) - \beta_1(a)\}\frac{\partial(xf)}{\partial x} . \tag{2.4}$$

It is well known (see, for example, Cox and Miller (1965, p. 226)) that if Z(0) is given, the diffusion process Z(t) has mean and variance

$$E[Z(t)] = Z(0)\exp[-\{\beta_2(a) - \beta_1(a)\}t] \tag{2.5}$$

$$\text{var}[Z(t)] = \alpha(a)\{\beta_2(a) - \beta_1(a)\}^{-1}\{1 - \exp[-2\{\beta_2(a) - \beta_1(a)\}t]\}. \tag{2.6}$$

Iglehart considered two special cases of this result, viz., the many server queue and the repairman problem. However, it is true that virtually all birth and death processes which arise in practice as models for physical phenomena converge (when transformed in the above sense) to Ornstein-Uhlenbeck processes. Thus we have a central limit theorem for birth and death processes.

In order to understand why the above theorem holds, we may proceed as follows. By the law of conditional probability

$$E[e^{i\theta Z_N(t+\delta t)}] = E[\sum_{i=1}^{3} E[e^{i\theta Z_N(t+\delta t)}|A_i, Z_N(t)]\Pr[A_i|Z_N(t)]], \tag{2.7}$$

where A_1, A_2 and A_3 are the events "one or more births in (t,t+δt)", "one or more deaths in (t,t+δt)" and "no change in (t,t+δt)", respectively. Using equations (2.1) and (2.3), equation (2.7) becomes

$$E[e^{i\theta Z_N(t+\delta t)}] = E[e^{i\theta Z_N(t)}] + (e^{i\theta/\sqrt{N}}-1)E[\lambda_N\{X_N(t)\}e^{i\theta Z_N(t)}]\delta t + (e^{-i\theta/\sqrt{N}}-1)E[\mu_N\{X_N(t)\}e^{i\theta Z_N(t)}]\delta t + o(\delta t),$$

whence, letting $t \to 0$, we derive

$$\frac{\partial}{\partial t}E[e^{i\theta Z_N(t)}] = (e^{i\theta/\sqrt{N}}-1)E[\lambda_N\{X_N(t)\}e^{i\theta Z_N(t)}] + (e^{-i\theta/\sqrt{N}}-1)E[\mu_N\{X_N(t)\}e^{i\theta Z_N(t)}]. \tag{2.8}$$

Now using equation (2.3) and the asymptotic formulas (2.2), expanding and collecting terms, equation (2.8) yields

$$\frac{\partial}{\partial t}\phi_N(\theta,t) = -\alpha(a)\theta^2\phi_N(\theta,t)$$
$$+ \{\beta_1(a) - \beta_2(a)\theta\frac{\partial}{\partial\theta}\phi_N(\theta,t) + 0(\frac{1}{\sqrt{N}}), \qquad (2.9)$$

where

$$\phi_N(\theta,t) = E[e^{i\theta Z_N(t)}]. \qquad (2.10)$$

Provided $\phi_N(\theta,t)$ converges to a limit, $\phi(\theta,t)$, say, as $N \to \infty$, this limit satisfies the partial differential equation

$$\frac{\partial\phi}{\partial t} = -\alpha(a)\theta^2\phi + \{\beta_1(a) - \beta_2(a)\}\theta\,\frac{\partial\phi}{\partial\theta}\,. \qquad (2.11)$$

Taking Fourier transforms in equation (2.11), we obtain (2.4). Thus it is plausible that $Z_N(t)$ should converge to the Ornstein-Uhlenbeck process with forward Kolmogorov equation (2.4).

There are a number of generalizations which immediately come to mind. In particular, it would be useful to have asymptotic results for multivariate birth and death processes, or *Markov population processes* as they are called (Kingman (1969)). Models for some more general birth and death processes in which the arrival (and/or departure) rates are time-dependent are also of interest in describing such phenomena as rush-hour traffic. Other processes not covered by the above theorem are transient birth and death processes, e.g. the simple epidemic. Unfortunately, we know of no general* results similar to Stone's which could be used to guarantee convergence to multivariate or temporally inhomogeneous diffusion processes for the above situations. However it is not difficult to generalize the heuristic argument

**Some particular results are known. Schach (1971), for example, has recently proved weak convergence of a sequence of multivariate Ehrenfest processes to the multivariate Ornstein-Uhlenbeck process.*

leading to equation (2.11) to these processes, and we will proceed in this way for a selection of models of scientific interest in subsequent sections.

We conclude this section with three examples.

(a) Generalized Ehrenfest processes: Taking, for λ, μ, $r > 0$,

$$\lambda(n) = \lambda(N - n)^r, \; \mu(n) = \mu n^r, \; (0 \leq X(t) \leq N) \tag{2.12}$$

we have Schach's (1970) generalization of Prendiville's (1949) Ehrenfest model, in which $r = 1$. Simple explicit formulas for the distribution of $X(t)$ are obtainable for the Ehrenfest process, but not for the general case $r \neq 1$. This model was suggested to simulate lane changing in a two-lane unidirectional motorway.

If we put

$$\lambda_N(n) = N^{1-r}\lambda(n), \; \mu_N(n) = N^{1-r}\mu(n), \tag{2.13}$$

(which simply amounts to a speeding-up or slowing-down of time, according as $r < 1$ or $r > 1$) the expansions (2.2) yield $\alpha(a) = \lambda(1-a)^r = \mu a^r$, $\beta_1(a) = -\lambda r(1-a)^{r-1}$ and $\beta_2(a) = \mu r a^{r-1}$. It follows that

$$a = 1/\{1 + (\mu/\lambda)^{1/r}\}. \tag{2.14}$$

Consequently $Z_N(t)$ converges weakly to the Ornstein-Uhlenbeck process $Z(t)$, which, using equation (2.5) and (2.6), has mean

$$Z(0)\exp\{-\mu r a^{r-1}(1-a)^{-1}t\}$$

and variance

$$(1-a)a^{-3}[1 - \exp\{-2\mu r a^{r-1}(1-a)^{-1}t\}]$$

Taking the limit as $t \to \infty$, we find that $X_N(t)$ is asymptotically normal with mean Na and variance $N(1-a)/a^3$.

(b) Stochastic logistic process: Prendiville (1949) suggested the Ehrenfest model for population growth because it is not possible to obtain exact results for the stochastic logistic process, in which, for λ, μ, $c > 0$

$$\lambda(n) = \lambda(c+n)(N-n), \quad \mu(n) = \mu n(c+n). \tag{2.15}$$

Here $X(t)$ is the amount by which the population exceeds its lower limit c, the upper limit being $c + N$. Putting

$$\lambda_N(n) = \lambda(\gamma N+n)(1-n/N), \quad \mu_N(n) = \mu n(\gamma+n/N), \tag{2.16}$$

(2.2) require that $\alpha(a) = \lambda(\gamma+a)(1-a) = \mu a(\gamma+a)$, $\beta_1(a) = \lambda(1-\gamma-2a)$ and $\beta_2(a) = \mu(\gamma+2a)$, whence

$$a = \lambda/(\lambda+\mu). \tag{2.17}$$

Thus the limiting Ornstein-Uhlenbeck process has mean $Z(0)\exp[-\{\lambda + (\lambda+\mu)\gamma\}t]$ and variance $\lambda\mu(\lambda+\mu)^{-2}\{1 - \exp[-2\{\lambda + (\lambda+\mu)\gamma\}t]\}$.

(c) Birth, death and migration process: For our next example we consider a population in which

$$\lambda(n) = \lambda n + \kappa, \quad \mu(n) = \mu n + \nu, \tag{2.18}$$

where λ, μ, κ, $\nu > 0$. The parameters λ and μ represent the birth and death rates per individual, while κ and ν are the immigration and emigration rates, respectively. In the case of no emigration ($\nu = 0$) it is possible to obtain the distribution of $X(t)$ exactly, and Kendall (1949) has given it. Even in this case the formula for the distribution is fairly complicated, and no exact formula at all is available if $\nu \neq 0$. Thus it is useful to have an asymptotic result.

In the previous examples, $X(t)$ took values on a bounded subset of the integers, and the parameter N arose in a natural way, as the size of this subset. In the present case, however, there is no such natural parameter, and various possibilities exist.

To begin with, let us try putting $\lambda_N(r) = \lambda(n)$, $\mu_N(n) = \mu(n)$. Then from equations (2.2), we must have $\alpha(a) = \lambda a = \mu a$, $\beta_1(a) = \lambda$, $\beta_2(a) = \mu$, so that $\lambda = \mu$ and κ and ν disappear in the limit. A more useful limit results if we put

$$\lambda_N(n) = \lambda n + \kappa N, \quad \mu_N(n) = \mu n + \nu N, \tag{2.19}$$

with $X_N(t)$ having initial state $X_N(0) = aN + Z(0)\sqrt{N}$. Thus the overall migration rate is of the same order of magnitude as the population size. Equations (2.2) and (2.19) yield $\alpha(a) = \lambda a + \kappa = \mu a + \nu$, $\beta_1(a) = \lambda$, $\beta_2(a) = \mu$, so that

$$a = \frac{\kappa-\nu}{\mu-\lambda} . \tag{2.20}$$

Since $a > 0$, it follows that either (i) $\lambda > \mu$ and $\nu > \kappa$, or (ii) $\lambda < \mu$ and $\nu < \kappa$. In the former case $\beta_2(a) - \beta_1(a)$ is negative, so that the limiting Ornstein-Uhlenbeck process has a drift away from the origin, and, using equations (2.5) and (2.6), the moments are

$$\begin{aligned} E[Z(t)] &= Z(0)e^{(\lambda-\mu)t}, \\ \mathrm{var}[Z(t)] &= \frac{\lambda\nu-\kappa\mu}{(\lambda-\mu)^2}\{e^{2(\lambda-\mu)t} - 1\}. \end{aligned} \tag{2.21}$$

In case (ii) $\beta_2(a) - \beta_1(a)$ is positive, and the limiting Ornstein-Uhlenbeck process is stationary. The formulas (2.21) remain valid in this case. The critical case $\lambda = \mu$, $\kappa = \nu$ yields a Wiener process, and a is determined by the initial condition, i.e.

$$a = \lim_{N\to\infty} \frac{X_N(0)}{N} \qquad (Z(0) = 0) \tag{2.22}$$

In this case $X_N(t)$ is asymptotically normal with mean aN and variance $2\lambda aN$.

It is possible to obtain a subtler limit when $\lambda = \mu$ by using the rescaling

$$\lambda_N(n) = \lambda n + \kappa\sqrt{N}, \quad \mu_N(n) = \lambda n + \nu\sqrt{N} \tag{2.23}$$

instead of equations (2.19). The asymptotic expansions (2.2) are not compatible with equations (2.23), but we find, instead

$$\begin{aligned} \lambda_N(aN + x\sqrt{N}) &= \lambda aN + (\lambda x+\kappa)\sqrt{N} + O(1), \\ \mu_N(aN + x\sqrt{N}) &= \lambda aN + (\lambda x+\nu)\sqrt{N} + O(1). \end{aligned} \tag{2.24}$$

Assuming, as before, that $Z_N(t)$ is given by equation (2.3), Iglehart's statement of Stone's result is sufficiently general to guarantee the weak convergence of $Z_N(t)$ to a diffusion process whenever the initial conditions converge. In this case the diffusion is a Wiener process with constant drift, characterized by the forward Kolmogorov equation

$$\frac{\partial f}{\partial x} = \lambda a \frac{\partial^2 f}{\partial x^2} - (\kappa-\nu)\frac{\partial f}{\partial x} . \tag{2.25}$$

As in the previous case, a is determined by the initial condition (2.22). The drift is upward or downward according as $\kappa > \nu$ or $\kappa < \mu$, and the conditional distribution of $Z(t)$ given $Z(0)$ was derived by Chandrasekhar (1943). It is also given by Cox and Miller (1965, p. 224).

The linear growth birth and death process admits yet another diffusion limit, if we rescale $\lambda(n)$ and $\mu(n)$ in the right way, and set $Z_N(t) = X_N(t)/N$, instead of (2.3). This possibility was suggested by Bailey (1964, p. 205), and yields a Gamma (rather than Gaussian) distribution in the limit. We will return to it in section 7. It should be noted that the type of limit, if any, which is useful depends on the physical phenomena being investigated.

3. MULTIVARIATE MARKOV PROCESSES

The multivariate generalization of a birth and death process is called a *Markov population process*, that is, a continuous time, non-negative integer component vector valued, Markov process X(t), with transition probabilities $\Pr[\underset{\sim}{X}(t+\delta t) - \underset{\sim}{X}(t) = i\underset{\sim}{e}_k - j\underset{\sim}{e}_\ell | \underset{\sim}{X}(t) = \underset{\sim}{n}]$ $(i \geq 0,\ j \geq 0)$, given by

$$\begin{cases} \lambda_k(\underset{\sim}{n})\delta t + o(\delta t) & \text{if } i = 1,\ j = 0, \\ \mu_\ell(\underset{\sim}{n})\delta t + o(\delta t) & \text{if } i = 0,\ j = 1, \\ \gamma_{\ell k}(\underset{\sim}{n})\delta t + o(\delta t) & \text{if } i = 1,\ j = 1, \\ 1 - \{\sum_k \lambda_k(\underset{\sim}{n}) + \sum_\ell \mu_\ell(\underset{\sim}{n}) + \sum_{k,\ell}\sum \gamma_{k\ell}(\underset{\sim}{n})\}\delta t + o(\delta t) & \text{if } i = j = 0, \\ o(\delta t), \text{ otherwise.} \end{cases} \tag{3.1}$$

In (3.1), $\underset{\sim}{e}_k$ denotes the vector with unity in position k and zeros elsewhere. Equilibrium distributions as $t \to \infty$ have been obtained for certain classes of Markov population processes by Kingman (1969).

The continuous sample path (diffusion) analogues to ergodic Markov population processes are multivariate Ornstein-Uhlenbeck processes. The basic theory of these processes is given by Schach (1971). To summarize, the joint characteristic function

$$\phi(\underset{\sim}{\theta}) = E[e^{i\underset{\sim}{\theta}'\underset{\sim}{Z}(t)}] \tag{3.2}$$

of a multivariate Ornstein-Uhlenbeck process $\underset{\sim}{Z}(t)$ satisfies the partial differential equation

$$\frac{\partial\phi}{\partial t} = \frac{1}{2}\underset{\sim}{\theta}'(\underset{\sim}{B}'\underset{\sim}{C} + \underset{\sim}{C}\underset{\sim}{B})\underset{\sim}{\theta}\phi + \underset{\sim}{\theta}'\underset{\sim}{B}\frac{\partial\phi}{\partial\theta}, \tag{3.3}$$

where $\underset{\sim}{C}$ is a covariance matrix. The mean vector and covariance matrix of $\underset{\sim}{Z}(t)$ are given by

$$E[\underset{\sim}{Z}(t)|\underset{\sim}{Z}(0)] = e^{\underset{\sim}{B}'t}\underset{\sim}{Z}(0) \tag{3.4}$$

$$E[\{\underset{\sim}{Z}(t) - e^{\underset{\sim}{B}'t}\underset{\sim}{Z}(0)\}\{\underset{\sim}{Z}(t) - e^{\underset{\sim}{B}'t}\underset{\sim}{Z}(0)\}']$$

$$= \underset{\sim}{C} - e^{\underset{\sim}{B}'t}\underset{\sim}{C}e^{\underset{\sim}{B}t}. \tag{3.5}$$

Equations (3.3), (3.4) and (3.5) generalize equations (2.11), (2.5) and (2.6), respectively.

The ojbect of the next three sections is to obtain asymptotic results for some particular Markov population models which are of scientific interest. The heuristic argument leading to equation (2.11) will be employed to obtain the form of the limiting process. While there seem to be no theorems similar to Stone's which can be used to establish such results there seems little doubt of their validity for the models investigated. We begin with a model, introduced by Kendall (1949) and studied further by Goodman (1953), which describes the growth of a bivariate population of males and females.

4. PROBLEM OF THE TWO SEXES

In the problem of the two sexes it is supposed that $\underset{\sim}{X}(t)$ is a two dimensional vector, whose components are the respective numbers of females and males in the population at time t. The transition probabilities are given by

$$\begin{cases} \lambda_1(\underset{\sim}{n}) = \lambda_2(\underset{\sim}{n}) = \frac{1}{2}\Lambda(\underset{\sim}{n}) \\ \mu_1(\underset{\sim}{n}) = \mu_1 n_1, \; \mu_2(\underset{\sim}{n}) = \mu_2 n_2, \\ \gamma_{12}(n) = \gamma_{21}(n) \equiv 0. \end{cases} \tag{4.1}$$

Thus it is assumed that there is a constant death rate μ_1 per individual female, and μ_2 for each male, and a birth rate which depends on the numbers of males and females in an as yet unspecified way. Each new individual is equally likely to be male or female.

Suppose now that we consider the sequence of transformed processes $\underset{\sim}{Z}_N(t)$, where

$$\underset{\sim}{Z}_N(t) = \{\underset{\sim}{X}_N(t) - \underset{\sim}{a}N\}/\sqrt{N}, \tag{4.2}$$

and $\underset{\sim}{a} > \underset{\sim}{0}$ is a two dimensional column vector, to be determined. The dependence $\underset{\sim}{X}(t) = \underset{\sim}{X}_N(t)$ arises through the birth rate $\Lambda(\underset{\sim}{n}) = \Lambda_N(\underset{\sim}{n})$, and we assume that

$$\Lambda_N(\underset{\sim}{a}N + \underset{\sim}{x}\sqrt{N}) = N\alpha(\underset{\sim}{a}) + \sqrt{N}\underset{\sim}{x}'\underset{\sim}{\beta}(\underset{\sim}{a}) + O(1), \tag{4.3}$$

where $\underset{\sim}{x}$ and $\underset{\sim}{\beta}(\underset{\sim}{a})$ are column vectors of dimension 2.

Proceeding with the heuristic derivation, we have, using conditional expectations

$$E[e^{i\underset{\sim}{\theta}'Z_N(t+\delta t)}] = E[\sum_{t=1}^{5} E[e^{i\underset{\sim}{\theta}'\underset{\sim}{Z}_N(t+\delta t)} | A_i; Z_N(t)] \Pr[A_i | Z_N(t)]] \tag{4.4}$$

where A_1, A_2, A_3, A_4 are the events "birth of a female", "birth of a male", "death of a female", and "death of a male", respectively, in the interval $(t, t+\delta t)$, and A_5 is the complement of the union of A_1, A_2, A_3 and A_4. Using equations (4.1) and (4.2) and taking the limit as $\delta t \to 0$, from (4.4) we derive

$$\frac{\partial}{\partial t}E[e^{i\underset{\sim}{\theta}'\underset{\sim}{Z}_N(t)}] = \frac{1}{2}(e^{i\underset{\sim}{\theta}_1/\sqrt{N}} + e^{i\underset{\sim}{\theta}_2/\sqrt{N}} - 2)E[\Lambda_N\{\underset{\sim}{X}_N(t)\}e^{i\underset{\sim}{\theta}'\underset{\sim}{Z}_N(t)}] + (\mu_1 \frac{\partial}{\partial\theta_1} + \mu_2 \frac{\partial}{\partial\theta_2})E[e^{i\underset{\sim}{\theta}'\underset{\sim}{Z}_N(t)}]. \tag{4.5}$$

Now using the expansion (4.3), expanding and collecting terms, equation (4.5) may be written as

$$\frac{\partial}{\partial t}\phi_N = i\sqrt{N}\{\tfrac{1}{2}(\theta_1+\theta_2)\alpha(\underset{\sim}{a}) - \mu_1 a_1\theta_1 - \mu_2 a_2\theta_2\}\phi_N$$

$$+ \{\tfrac{1}{2}(\theta_1+\theta_2)\beta_1(\underset{\sim}{a})-\mu_1\theta_1\}\frac{\partial}{\partial\theta_1}\phi_N+\{\tfrac{1}{2}(\theta_1+\theta_2)\beta_2(\underset{\sim}{a})-\mu_2\theta_2\}\frac{\partial}{\partial\theta_2}\phi_N$$

$$- \tfrac{1}{2}\{\tfrac{1}{2}(\theta_1^2+\theta_2^2)\alpha(\underset{\sim}{a})+a_1\mu_1\theta_1^2+a_2\mu_2\theta_2^2\}\phi_N + 0(\tfrac{1}{\sqrt{N}}),$$

where $\phi_N = \phi_N(\underset{\sim}{\theta},t) = E[\exp\{i\underset{\sim}{\theta}'\underset{\sim}{Z}_N(t)\}]$. Thus provided

$$\mu_1 a_1 = \mu_2 a_2 = \frac{1}{2}\alpha(\underset{\sim}{a}) \tag{4.7}$$

and assuming convergence, $\phi_N \to \phi$ as $N \to \infty$, where

$$\frac{\partial\phi}{\partial t} = -\frac{1}{2}\alpha(\underset{\sim}{a})\underset{\sim}{\theta}'\underset{\sim}{\theta}\phi + \underset{\sim}{\theta}'\underset{\sim}{B}\frac{\partial\phi}{\partial\underset{\sim}{\theta}}, \tag{4.8}$$

and

$$\underset{\sim}{B} = \begin{pmatrix} \frac{1}{2}\beta_1(\underset{\sim}{a}) - \mu_1 & \frac{1}{2}\beta_2(\underset{\sim}{a}) \\ \frac{1}{2}\beta_1(\underset{\sim}{a}) & \frac{1}{2}\beta_2(\underset{\sim}{a}) - \mu_2 \end{pmatrix} \tag{4.9}$$

Comparing equations (3.3) and (4.8), we must have

$$\underset{\sim}{B}'\underset{\sim}{C} + \underset{\sim}{C}\underset{\sim}{B} = -\alpha(\underset{\sim}{a})\underset{\sim}{I}. \tag{4.10}$$

Solving (4.10) for $\underset{\sim}{C}$, we find (see, for example, Bellman (1970, p. 179))

$$\underset{\sim}{C} = \alpha(\underset{\sim}{a})\int_0^\infty e^{(\underset{\sim}{B}'+\underset{\sim}{B})t}dt = \alpha(\underset{\sim}{a})(-\underset{\sim}{B}-\underset{\sim}{B})^{-1}, \tag{4.11}$$

provided the integral exists. The integral exists provided $\underset{\sim}{B}$ is nonsingular and its eigenvalues do not sum to zero (Bellman, p. 239).

Let us consider some particular forms for the birth rate $\Lambda(\underset{\sim}{n})$.

(a) Root product birth rate: Kendall (1949) suggested a number of functional forms for Λ, including the relation

$$\Lambda(n_1,n_2) = 2\lambda\sqrt{(n_1 n_2)}. \tag{4.12}$$

Expanding this as in (4.3), we find $\alpha(\underset{\sim}{a}) = 2\lambda\sqrt{(a_1 a_2)}$, $\beta_1(\underset{\sim}{a}) = \lambda\sqrt{(a_2/a_1)}$, $\beta_2(\underset{\sim}{a}) = \lambda\sqrt{(a_1/a_2)}$, so that equation (4.7) requires that

$$\mu_1 a_1 = \mu_2 a_2 = \lambda\sqrt{(a_1 a_2)}. \tag{4.13}$$

Equation (4.13) implies $\lambda = \sqrt{(\mu_1\mu_2)}$, in which case

$$\underset{\sim}{B} = \begin{pmatrix} -\frac{1}{2}\mu_1 & \frac{1}{2}\mu_2 \\ \frac{1}{2}\mu_1 & -\frac{1}{2}\mu_2 \end{pmatrix}, \tag{4.14}$$

so that $\det(\underset{\sim}{B}) = 0$ and the integral in (4.11) diverges. However if we make the transformation $\underset{\sim}{U}(t) = \underset{\sim}{A}\underset{\sim}{Z}(t)$, and put $\psi = E[\exp\{i\underset{\sim}{\theta}'\underset{\sim}{U}(t)\}]$, from equation (4.8) we derive

$$\frac{\partial\psi}{\partial t} = -\mu_1 a_1 \underset{\sim}{\theta}'\underset{\sim}{A}\underset{\sim}{A}'\underset{\sim}{\theta}\psi + \underset{\sim}{\theta}'\underset{\sim}{A}\underset{\sim}{B}\underset{\sim}{A}^{-1}\frac{\partial\psi}{\partial\underset{\sim}{\theta}}. \tag{4.15}$$

Now choosing $\underset{\sim}{A}$ in such a way that $\underset{\sim}{A}\underset{\sim}{B}\underset{\sim}{A}^{-1}$ is a diagonal matrix, equation (4.15) becomes the (Fourier transformed) forward Kolmogorov equation of a bivariate process, one component of which is an Ornstein-Uhlenbeck process, the other a Wiener process.

An appropriate choice of $\underset{\sim}{A}$ is

$$\underset{\sim}{A} = \begin{pmatrix} 1 & -\mu_2/\mu_1 \\ 1 & 1 \end{pmatrix}, \tag{4.16}$$

from which equation (4.15) becomes

$$\frac{\partial\psi}{\partial t} = -\mu_1 a_1\{(1+\mu_2^2/\mu_1^2)\theta_1^2 + 2(1-\mu_2/\mu_1)\theta_1\theta_2 + 2\theta_2^2\}\psi$$
$$-\frac{1}{2}(\mu_1+\mu_2)\theta_1\frac{\partial\psi}{\partial\theta_1}. \tag{4.17}$$

Equation (4.17) is easily solved; we get

$$\psi = \exp\{i\theta_1 U_1(0)e^{-\frac{1}{2}(\mu_1+\mu_2)t}$$
$$-\frac{1}{2}a_1[\theta_1^2\sigma_1^2(t) + 2\theta_1\theta_2\gamma(t) + \theta_2^2\sigma_2^2(t)]$$

where

$$\sigma_1^2(t) = 2(1+\mu_2^2/\mu_1^2)(1+\mu_2/\mu_1)^{-1}(1-e^{-(\mu_1+\mu_2)t}),$$

$$\sigma_2^2(t) = 4\mu_1 t \tag{4.19}$$

$$\gamma(t) = 4(\mu_1-\mu_2)(\mu_1+\mu_2)^{-1}(1-e^{-(\mu_1+\mu_2)t/2}).$$

Thus $U_1(t) = Z_1(t) - (\mu_2/\mu_1)Z_2(t)$ and $U_2(t) = Z_1(t) + Z_2(t)$ have correlation coefficient

$$\text{corr}[U_1(t),U_2(t)]$$
$$= \frac{\mu_1\mu_2}{\surd(2\mu_1^2+2\mu_2^2)} \cdot \frac{2\{1-e^{-(\mu_1+\mu_2)t/2}\}}{\surd\{t-te^{-(\mu_1+\mu_2)t}\}\surd(\mu_1+\mu_2)}, \tag{4.20}$$

which is zero if $\mu_1 = \mu_2$, and tends to zero in any case as $t \to \infty$.

It may be concluded that if males and females have the same death rate, the excess of females over males becomes independent of the population size as this quantity tends to infinity.

(b) Female dominance: Another model suggested by Kendall in his 1949 paper is that in which

$$\Lambda(n_1,n_2) = 2\lambda n_1. \tag{4.21}$$

This situation is usually referred to as "female dominance", since the birth rate is independent of the size of the male population. Using the expansion (4.3), we find $\alpha(\underset{\sim}{a}) = 2\lambda a_1$, $\beta_1(\underset{\sim}{a}) = 2\lambda$, $\beta_2(\underset{\sim}{a}) = 0$, whence equations (4.7) imply

$$\mu_1 a_1 = \mu_2 a_2 = \lambda a_1. \tag{4.22}$$

Thus convergence to a stationary bivariate Gaussian process is possible only if $\lambda = \mu_1$. Equation (4.19) then becomes

$$\underset{\sim}{B} = \begin{pmatrix} 0 & 0 \\ \mu_1 & -\mu_2 \end{pmatrix}, \tag{4.23}$$

so that $\underset{\sim}{B}$ is again singular. In this case we may transform Z(t) with the operator

$$\underset{\sim}{A} = \begin{pmatrix} 1 & -\mu_2/\mu_1 \\ 1 & 0 \end{pmatrix}, \tag{4.24}$$

and equation (4.15) may be written as

$$\frac{\partial\psi}{\partial t} = -\mu_1 a_1\{(1+\mu_2^2/\mu_1^2)\theta_1^2 + 2\theta_1\theta_2 + \theta_2^2\}\psi - \mu_2\theta_1\frac{\partial\psi}{\partial\theta_1}. \tag{4.25}$$

From equation (4.25) we may calculate the moments of $\{Z_1(t) - (\mu_2/\mu_1)Z_2(t), Z_1(t)\}$; the correlation coefficient is found to be

$$\rho(t) = \frac{\mu_1}{\sqrt{(\mu_1^2+\mu_2^2)}}\frac{\sqrt{2}\{1-e^{-\mu_2 t}\}}{\sqrt{\{\mu_2 t-\mu_2 te^{-2\mu_2 t}\}}}, \tag{4.26}$$

which tends to zero as $t \to \infty$, but is never zero for any values of μ_1 and μ_2 for $t < \infty$.

Thus the case of female dominance is similar qualitatively to the case of intermediate dominance. In each case the population may be regarded as consisting in the diffusion limit, of an Ornstein-Uhlenbeck process, representing the excess of females over "weighted" males, superimposed upon a Wiener process representing the magnitude of the population, and in the limit as $t \to \infty$, these components become independent.

One could go on, and consider (as Kendall did) married and single people separately. This would give rise to a trivariate process. One could go further and introduce social groups with differing birth and death rates. However the crucial feature in the case of human (indeed most animal) populations is the dependence of fertility and to a lesser extent mortality on age, and this is a major drawback in the above type of model for population growth. It would probably be better to test the validity and relevance of these models for particular scientific applications before developing the theory much further.

5. MULTIVARIATE GENERALIZED EHRENFEST PROCESS

One obtains a multivariate generalization of the Ehrenfest process if in (3.1) one puts

$$\begin{aligned} \lambda_k(\underset{\sim}{n}) &= \mu_k(\underset{\sim}{n}) = 0, \quad k = 1,2,\ldots K, \\ \gamma_{k\ell}(\underset{\sim}{n}) &= \lambda_{k\ell} n_k^r (N-n_\ell)^s, \quad k,\ell = 1,2,\ldots K, \end{aligned} \tag{5.1}$$

where $r \geq 0$, $s \geq 0$. Thus $\underset{\sim}{X}(t)$ is a K-variate Markov population process, degenerate since $\underset{\sim}{e}'\underset{\sim}{X}(t) = N$. ($\underset{\sim}{e}$ is the column vector of ones.) To obtain a valid diffusion limit, it is necessary to replace $\lambda_{k\ell}$ by $N^{1-r-s}\alpha_{k\ell}$, so that, from (5.1),

$$\gamma_{k\ell}(\underset{\sim}{a}N+\underset{\sim}{x}\sqrt{N}) = N\alpha_{k\ell}a_k^r(1-a_\ell)^s + \sqrt{N}\{\frac{rx_k}{a_k} - \frac{sx_\ell}{1-a_\ell}\} + O(1) \tag{5.2}$$

For the analogue to equation (4.6) we now derive

$$\begin{aligned} \frac{\partial}{\partial t}\phi_N &= i\sqrt{N}\sum_{k,\ell}(\theta_\ell-\theta_k)\alpha_{k\ell}a_k^r(1-a_\ell)^s\phi_N \\ &+ \sum_{k,\ell}(\theta_\ell-\theta_k)\alpha_{k\ell}a_k^r(1-a_\ell)^s\{\frac{r}{a_k}\frac{\partial}{\partial\theta_k}\phi_N - \frac{s}{1-a_\ell}\frac{\partial}{\partial\theta_\ell}\phi_N\} \\ &- \frac{1}{2}\sum_{k,\ell}(\theta_k-\theta_\ell)^2\alpha_{k\ell}a_k^r(1-a_\ell)^s\phi_N + o(1) \end{aligned} \tag{5.3}$$

Thus if ϕ_N is to converge the coefficient of $\sqrt{N}$ in the right-hand side of equation (5.3) must vanish for all $\underset{\sim}{\theta}$. This means that

$$(1-a_k)^s\sum_{\ell=1}^{K}\alpha_{\ell k}a_\ell^r = a_k^r\sum_{\ell=1}^{K}\alpha_{k\ell}(1-a_\ell)^s, \quad k = 1,2,\ldots,K. \tag{5.4}$$

(It should be noted that $\sum_{\ell=1}^{K} a_\ell = 1$, $a_\ell \geq 0$.) We assume that it is possible to find $\underset{\sim}{a}$ such that equation (5.4) is valid. In this case we expect ϕ_N to tend to ϕ as $N \to \infty$, where $\phi = \phi(\underset{\sim}{\theta},t)$ satisfies

$$\frac{\partial\phi}{\partial t} = -\frac{1}{2}\sum_{k,\ell}(\theta_k-\theta_\ell)^2\alpha_{k\ell}a_k^r(1-a_\ell)^s\phi$$
$$+\sum_{k,\ell}(\theta_\ell-\theta_k)\alpha_{k\ell}a_k^r(1-a_\ell)^s\{\frac{r}{a_k}\frac{\partial\phi}{\partial\theta_k} - \frac{s}{1-a_\ell}\frac{\partial\phi}{\partial\phi_\ell}\} . \tag{5.5}$$

For purposes of illustration, we consider two examples in detail.

(a) Whittle migration process: If we take $s = 0$, the transition rate from component k to component ℓ is independent of the size of component ℓ, and the model is one of a class considered by Whittle (1967). Let us put

$$\sum_{\ell=1}^{K}\alpha_{k\ell} = \alpha_k, \quad p_{k\ell} = \alpha_{k\ell}/\alpha_k. \tag{5.6}$$

Then equations (5.4) become

$$\sum_{\ell=1}^{K} p_{\ell k}\alpha_\ell a_\ell^r = a_k^r\alpha_k, \quad k = 1,2,\ldots,K. \tag{5.7}$$

If $\underset{\sim}{P} = (p_{k\ell})$ is ergodic (i.e., only one communicating class) then equations (5.7) have a unique non-negative solution up to a constant factor, which is determined by $\Sigma a_k = 1$. (In fact $(\alpha_1 a_1^r,\ldots,\alpha_k a_k^r)$ is the unique left eigenvector of $\underset{\sim}{P}$ corresponding to the eigenvalue 1.)

When $s = 0$, equation (5.5) may be written in the matrix form

$$\frac{\partial\phi}{\partial t} = -\frac{1}{2}\underset{\sim}{\theta}'\{\underset{\sim}{A}^r\underset{\sim}{\Lambda}(\underset{\sim}{I}-\underset{\sim}{P}) + (\underset{\sim}{I}-\underset{\sim}{P}')\underset{\sim}{\Lambda}\underset{\sim}{A}^r\}\underset{\sim}{\theta}\phi$$
$$- r\underset{\sim}{\theta}(\underset{\sim}{I}-\underset{\sim}{P}')\underset{\sim}{A}^{r-1}\underset{\sim}{\Lambda}\frac{\partial\phi}{\partial\underset{\sim}{\theta}} , \tag{5.8}$$

where $\underset{\sim}{A}$ and $\underset{\sim}{\Lambda}$ are the diagonal matrices with elements a_k and α_k, respectively.

In order to obtain explicit formulas for the variance-covariance matrix of the process, we need the stationary variance-covariance matrix, denoted by $\underset{\sim}{C}$ in equations (3.3) and (3.5). Comparing equations (3.3) and (5.8), we have

$$\underset{\sim}{B}'\underset{\sim}{C} + \underset{\sim}{C}\underset{\sim}{B} = -\underset{\sim}{A}^r\underset{\sim}{\Lambda}(\underset{\sim}{I}-\underset{\sim}{P}) - (\underset{\sim}{I}-\underset{\sim}{P}')\underset{\sim}{\Lambda}\underset{\sim}{A}^r, \tag{5.9}$$

where $\underset{\sim}{B} = -r(\underset{\sim}{I}-\underset{\sim}{P})\underset{\sim}{A}^{r-1}\underset{\sim}{\Lambda}$. Solving equation (5.9) for $\underset{\sim}{C}$, we find (Bellman (1970, p. 179))

$$\underset{\sim}{C} = \int_0^\infty e^{\underset{\sim}{B}'t}\{\underset{\sim}{A}^r\underset{\sim}{\Lambda}(\underset{\sim}{I}-\underset{\sim}{P}) - (\underset{\sim}{I}-\underset{\sim}{P}')\underset{\sim}{\Lambda}\underset{\sim}{A}^r\}e^{\underset{\sim}{B}t}dt,$$

$$= -\{r^{-1}\underset{\sim}{G}'\underset{\sim}{A}\underset{\sim}{G} - \underset{\sim}{A}\}, \tag{5.10}$$

where $\underset{\sim}{G} = \lim_{t\to\infty} \exp\{r\underset{\sim}{A}^{r-1}\underset{\sim}{\Lambda}(\underset{\sim}{P}-\underset{\sim}{I})t\}$.

It remains to evaluate $\underset{\sim}{G}$. We note that $\underset{\sim}{Q}(t) = \exp\{(\underset{\sim}{P}-\underset{\sim}{I})t\}$ is the transition probability matrix of a pseudo-Poissonian Markov process with infinitesimal generator $\underset{\sim}{P} - \underset{\sim}{I}$ (Feller (1971, p. 345)). Such a process stays in each state for an exponential (with unit mean) length of time, and then jumps according to the discrete-time chain with matrix $\underset{\sim}{P}$. It is easy to see that $r\underset{\sim}{A}^{r-1}\underset{\sim}{\Lambda}(\underset{\sim}{P}-\underset{\sim}{I})$ is still an infinitesimal generator. (In other words, the rows of $r\underset{\sim}{A}^{r-1}\underset{\sim}{\Lambda}(\underset{\sim}{P}-\underset{\sim}{I})$ are intensities; they are non-negative except for diagonal elements, and the rows sum to zero.) Multiplication by the diagonal matrix $r\underset{\sim}{A}^{r-1}\underset{\sim}{\Lambda}$ affects the expected sojourn times in individual states. Thus

$$\overline{\underset{\sim}{Q}}(t) = \exp\{r\underset{\sim}{A}^{r-1}\underset{\sim}{\Lambda}(\underset{\sim}{P}-\underset{\sim}{I})t\}$$

is, for each t, a transition probability matrix, and the corresponding process is irreducible, since $\underset{\sim}{P}$ was assumed to be irreducible.

General Markov theory now tells us that $\overline{\underset{\sim}{Q}}(t)$ has a limit $\underset{\sim}{G}$ as $t \to \infty$, that the rows of this limit are identical, each equal to $\underset{\sim}{g}'$, say, and that $\underset{\sim}{g}'$ is determined uniquely by

$$\underset{\sim}{g}'\overline{\underset{\sim}{Q}}(t) = \underset{\sim}{g}'.$$

By the spectral mapping theorem this is equivalent to

$$\underset{\sim}{g}'r\underset{\sim}{A}^{r-1}\underset{\sim}{\Lambda}(\underset{\sim}{P}-\underset{\sim}{I}) = \underset{\sim}{0}'.$$

Now since the left eigenvector of $\underset{\sim}{P}$ corresponding to the eigenvalue 1 has the form $(\alpha_1 a_1^r, \ldots, \alpha_K a_K^r)$, it follows that $\underset{\sim}{g}' = (a_1, a_2, \ldots, a_K)$, and hence

$$\underset{\sim}{G} = \begin{pmatrix} a_1 & a_2 & \cdots & a_K \\ a_1 & a_2 & \cdots & a_K \\ \cdot & \cdot & & \cdot \\ \cdot & \cdot & & \cdot \\ a_1 & a_2 & \cdots & a_K \end{pmatrix} . \tag{5.11}$$

The purpose of this exercise was to compute the stationary covariance matrix $\underset{\sim}{C}$, given by equation (5.10). Using equations (5.10) and (5.11), we find

$$\underset{\sim}{C} = r^{-1}(\underset{\sim}{A} - \underset{\sim}{a}\underset{\sim}{a}'). \tag{5.12}$$

(b) Symmetric case: The analysis is greatly simplified if the transition rates are symmetric, i.e., $\alpha_{k\ell} = \alpha_{\ell k}$, for all k,ℓ. If we try a solution $a_\ell = a$, $\ell = 1,2,\ldots,K$, equation (5.4) becomes

$$(1-a)^s a^r \sum_{\ell=1}^{K} \alpha_{k\ell} = a^r (1-a)^s \sum_{\ell=1}^{K} \alpha_{k\ell}, \quad k = 1,2,\ldots,K,$$

which is obviously satisfied. Hence $a_k = 1/K$, $k = 1,2,\ldots,K$. The matrix form of equation (5.5) is now

$$\begin{aligned} \frac{\partial\phi}{\partial t} = &- \frac{1}{2} K^{-r-s}(K-1)^s \underset{\sim}{\theta}'\{\underset{\sim}{\Lambda}(\underset{\sim}{I}-\underset{\sim}{P}) + (\underset{\sim}{I}-\underset{\sim}{P}')\underset{\sim}{\Lambda}\}\underset{\sim}{\theta}\phi \\ &+ K^{1-r-s}(K-1)^{s-1}\{(K-1)r+s\}\underset{\sim}{\theta}'\underset{\sim}{\Lambda}(\underset{\sim}{P}-\underset{\sim}{I})\frac{\partial\phi}{\partial\underset{\sim}{\theta}} , \end{aligned} \tag{5.13}$$

where $\underset{\sim}{P}$ and $\underset{\sim}{\Lambda}$ are defined as before. The covariance matrix $\underset{\sim}{C}$ is easily calculated to be

$$\underset{\sim}{C} = \frac{K-1}{K^2\{(K-1)r+s\}} \{K\underset{\sim}{I} - \underset{\sim}{e}\underset{\sim}{e}'\}, \tag{5.14}$$

where $\underset{\sim}{e}$ is the vector of units. Thus, in the limiting distribution, as $t \to \infty$, the correlation between any two different components is $-1/(K-1)$. It seems at first sight surprising that such a simple result is obtained for quite general parameters $\alpha_{k\ell}$, subject only to the apparently mild symmetry condition $\alpha_{\ell k} = \alpha_{k\ell}$.

This condition is identical to the reversibility condition for a Markov population process (see Kingman (1969)).

6. GROWTH OF TWO CITIES

As a final example of a multivariate birth and death process we consider the growth and interraction of cities. Besides the usual assumptions for births and deaths within each city, it will be supposed that migration occurs as follows. Emigrations from a city occur in such a way that an inhabitant's decision to emigrate is independent of the size of the city of abode. Having decided to emigrate, the probability of choosing to live in a particular city is a linear (increasing) function of its size. Considering K such cities, whose sizes at time t are $\underset{\sim}{X}(t) = (X_1(t),\ldots,X_K(t))$, we are led to the transition probabilities (3.1) with

$$\begin{cases} \lambda_k(\underset{\sim}{n}) = \lambda_k n_k + \kappa_k, & k = 1,2,\ldots,K, \\ \mu_\ell(\underset{\sim}{n}) = \mu_\ell n_\ell, & \ell = 1,2,\ldots,K, \\ \gamma_{k\ell}(\underset{\sim}{n}) = \rho_{k\ell} n_k + \beta_{k\ell} n_k n_\ell, & k,\ell = 1,2,\ldots,K, \quad (k \neq \ell) \end{cases} \tag{6.1}$$

where $\lambda_k, \mu_k, \rho_{k\ell}, \beta_{k\ell} > 0$, for all k,ℓ.

A sequence of processes $\underset{\sim}{Z}_N(t)$, for which convergence to a K-variate Ornstein-Uhlenbeck process may be possible, is obtained by replacing κ_k and $\beta_{k\ell}$ by $N\kappa_k$ and $(1/N)\beta_{k\ell}$ respectively, so that $\underset{\sim}{X}(t) = \underset{\sim}{X}_N(t)$, and defining $\underset{\sim}{Z}_N(t)$ as in equation (4.2). To keep the algebra simple, let us consider the case when there are just two cities, so that K = 2. Then writing down the (Fourier transformed) forward Kolmogorov equation for $\underset{\sim}{Z}_N(t)$, and taking the limit as $N \to \infty$, we obtain

$$\frac{\partial\phi}{\partial t} = -\frac{1}{2}\underset{\sim}{\theta}'\underset{\sim}{D}\underset{\sim}{\theta}\phi + \underset{\sim}{\theta}'\underset{\sim}{B}\frac{\partial\phi}{\partial\underset{\sim}{\theta}} \tag{6.2}$$

where

$$\underset{\sim}{D} = \begin{pmatrix} \kappa_1+\rho_{21}a_2+(\lambda_1+\mu_1+\rho_{12})a_1+\sigma a_1a_2 & -\rho_{12}a_1-\rho_{21}a_2-\sigma a_1a_2 \\ -\rho_{12}a_1-\rho_{21}a_2-\sigma a_1a_2 & \kappa_2+\rho_{12}a_1+(\lambda_2+\mu_2+\rho_{21})a_2+\sigma a_1a_2 \end{pmatrix} \tag{6.3}$$

and

$$\underset{\sim}{B} = \begin{pmatrix} \lambda_1-\mu_1-\rho_{12}+\gamma a_2 & \rho_{21}+\gamma a_1 \\ \rho_{12}-\gamma a_2 & \lambda_2-\mu_2-\rho_{21}-\gamma a_1 \end{pmatrix} \tag{6.4}$$

(We have put $\sigma = \beta_{21} + \beta_{12}$, $\gamma = \beta_{21} - \beta_{12}$.) The constants a_1 and a_2 satisfy the equations

$$\begin{cases} (\rho_{12}-\lambda_1+\mu_1)a_1 - \gamma a_1a_2 - a_2\rho_{21} = \kappa_1 \\ (\rho_{21}-\lambda_2+\mu_2)a_2 + \gamma a_1a_2 - a_1\rho_{12} = \kappa_2. \end{cases} \tag{6.5}$$

Qualitatively, the bivariate process $\underset{\sim}{Z}(t)$ is similar to the linear growth birth and death process considered in section 2. For certain values of the parameters, a stationary bivariate Ornstein-Uhlenbeck process results; for other values the drift is away from the origin so that we have exponential growth. To study this behavior in greater detail let us assume, for simplicity, that $\beta_{21} = \beta_{12}$, so that $\gamma = 0$. Then from equations (6.5)

$$\begin{cases} a_1 = \dfrac{\kappa_2\rho_{21}+(\rho_{21}+\mu_2-\lambda_2)\kappa_1}{(\rho_{21}+\mu_2-\lambda_2)(A_2+\mu_1-\lambda_1)-\rho_{12}\rho_{21}} \\ a_2 = \dfrac{\kappa_1\rho_{12}+(\rho_{12}+\mu_1-\lambda_1)\kappa_2}{(\rho_{21}+\mu_2-\lambda_2)(\rho_{12}+\mu_1-\lambda_1)-\rho_{12}\rho_{21}} \end{cases}. \tag{6.6}$$

In order that a_1 and a_2 be finite and positive, it is thus sufficient that $\det(\underset{\sim}{B}) > 0$ and $\lambda_1 < \mu_1 + \rho_{12}$. In this case the process $\underset{\sim}{Z}(t)$ is a stationary Ornstein-Uhlenbeck process. To further simplify the algebra, consider the symmetric case, i.e., $\lambda_1 = \lambda_2 = \lambda$, $\mu_1 = \mu_2 = \mu$, $\kappa_1 = \kappa_2 = \kappa$, $\rho_{21} = \rho_{12} = \rho$. Then equations (6.6) reduce to

$$a_1 = a_2 = \frac{\kappa}{\mu-\lambda} \tag{6.7}$$

which is the same as for the univariate linear growth birth death and immigration process discussed in section 2. For this case equation (6.3) becomes

$$\underset{\sim}{D} = \frac{\kappa}{(\mu-\lambda)^2}\begin{pmatrix} 2(\mu+\rho)(\mu-\lambda)+\sigma\kappa & -2\rho(\mu-\lambda)-\sigma\kappa \\ -2\rho(\mu-\lambda)-\sigma\kappa & 2(\mu+\rho)(\mu-\lambda)+\sigma\kappa \end{pmatrix}. \tag{6.8}$$

The covariance matrix of the limiting distribution of $\underset{\sim}{Z}(t)$ (as $t \to \infty$) is thus

$$\underset{\sim}{C} = -\int_0^\infty e^{\underset{\sim}{B}'t}\underset{\sim}{D}e^{\underset{\sim}{B}t}dt = -\frac{1}{2}\underset{\sim}{B}^{-1}\underset{\sim}{D}.$$

Using equations (6.4) and (6.8) this yields, after a few calculations

$$\underset{\sim}{C} = \frac{\kappa}{2(\mu-\lambda)^2(\mu-\lambda-2\rho)}\begin{pmatrix} 2\mu\rho+\sigma\kappa+2(\mu-\lambda)(\rho+\mu) & 2\lambda\rho-\sigma\kappa \\ 2\lambda\rho-\sigma\kappa & 2\mu\rho+\sigma\kappa+2(\mu-\lambda)(\rho+\mu) \end{pmatrix}. \tag{6.9}$$

Thus the limiting joint distribution factorizes into independent components if $\frac{1}{2}\sigma\kappa = \lambda\rho$. Kingman (1969) obtained a much more general result, i.e., for the process $\underset{\sim}{X}(t)$ defined by (6.1), the components are independent whenever $\beta_{k\ell}\kappa_k = \lambda_k\rho_{k\ell}$, for all k,ℓ.

7. SOME RELATED RESULTS

In section 2 it was mentioned that diffusion processes other than the Ornstein-Uhlenbeck process may be obtained as limits of birth and death processes. In this section we discuss three such situations.

(a) Gamma process: Consider again the linear growth, birth, death and migration process. For a sequence $\{X_N(t)\}$ of such processes, take the parameters to be

$$\lambda_N(n) = N\lambda n + N\kappa, \quad \mu_N(n) = (N\lambda+\alpha)n + N\nu. \tag{7.1}$$

Thus the ratio of birth rate to the death rate (per individual) tends to unity as $N \to \infty$. Suppose now that we make the rescaling

$$Z_N(t) = X_N(t)/N, \tag{7.2}$$

instead of equation (2.3). Then the (Laplace-Stieltjes-transformed) forward Kolmogorov equation for $Z_N(t)$ is

$$\begin{aligned}\frac{\partial}{\partial t}\phi_N &= -N\{N\lambda(e^{-\theta/N}-1) + (N\lambda+\alpha)(e^{\theta/N}-1)\}\frac{\partial}{\partial\theta}\phi_N \\ &\quad + N\{\kappa(e^{-\theta/N}-1) + \nu(e^{\theta/N}-1)\}\phi_N,\end{aligned} \tag{7.3}$$

where $\phi_N = \phi_N(\theta,t) = E[\exp\{-\theta Z_N(t)\}]$. In the limit as $N \to \infty$, equation (7.3) becomes

$$\frac{\partial\phi}{\partial t} = -(\lambda\theta^2 + \alpha\theta)\frac{\partial\phi}{\partial\theta} - (\kappa-\nu)\theta\phi, \tag{7.4}$$

where it is assumed that the limit $\phi = \lim_{N\to\infty}\phi_N$ exists. Inverting equation (7.4), we obtain

$$\frac{\partial f}{\partial t} = \lambda\frac{\partial^2}{\partial x^2}(xf) - \frac{\partial}{\partial x}\{(\kappa-\nu-\alpha x)f\}. \tag{7.5}$$

Equation (7.5) is the forward Kolmogorov equation for a diffusion process $Z(t)$ with instantaneous acceleration $\lambda Z(t)$ and drift $\kappa-\nu-\alpha Z(t)$. Thus a stationary process can exist for large t only if $\kappa > \nu$ and $\alpha > 0$. If neither of these conditions is satisfied there is a tendency for the process to drift to infinity or to be absorbed at the origin.

The equilibrium distribution, obtained by putting $\partial f/\partial t = 0$ in (7.5) and solving, is

$$f(x) = \frac{\alpha/\lambda}{\Gamma\{(\kappa-\nu)/\lambda\}}\left(\frac{ax}{\lambda}\right)^{(\kappa-\nu-\lambda)/\lambda} e^{-(\alpha/\lambda)x}, \quad x \geq 0. \tag{7.6}$$

Thus the stationary distribution is a Gamma distribution with scale a/λ and $2(\kappa-\nu)/\lambda$ degrees of freedom. In fact it is possible to obtain the time-dependent (conditional) distribution of $Z(t)$, given $Z(0)$. Equation (7.4) may be solved by the method of characteristics to give

$$\phi(\theta,t) = \{1+\frac{\lambda}{\alpha}(1-e^{-\alpha t})\theta\}^{-(\kappa-\nu)/\lambda}\exp\left\{\frac{-\theta Z(0)e^{-\alpha t}}{1+\lambda\alpha^{-1}(1-e^{-\alpha t})\theta}\right\}. \quad (7.7)$$

Inverting this Laplace transform, we find (Erdelyi (1954, p. 197))

$$f(x,t) = \beta(t)e^{-\beta(t)x-Z(0)/\gamma(t)}\left\{\frac{\beta(t)\gamma(t)x}{Z(0)}\right\}^{r/2} I_r\left\{2\left(\frac{xZ(0)\beta(t)}{\gamma(t)}\right)^{\frac{1}{2}}\right\}, \quad (7.8)$$

where $\beta(t) = \alpha\lambda^{-1}(1-e^{-\alpha t})^{-1}$, $\gamma(t) = \alpha^{-1}\lambda e^{\alpha t}(1-e^{-\alpha t})$, $r = (\kappa-\nu-\lambda)/\lambda$, and $I_r(X)$ is a modified Bessel function of the first kind. The right-hand side of equation (7.8) is identical to a conditional probability density function obtained by Lampard (1968). In fact, if we assume that $Z(0)$ has the stationary distribution given by (7.6) (this is equivalent to assuming that the process is in equilibrium) then the joint probability density function of $Z(0)$ and $Z(t)$ is $f(x_1,x_2;t) = f(x_2,t)f(x_1)$. Using equations (7.6) and (7.8) and simplifying the result, this becomes

$$f(x_1,x_2;t) \quad (7.9)$$
$$= \left(\frac{\alpha}{\lambda}\right)^{r+2}\left\{\frac{\tau^{-r}}{1-\tau^2}\right\}\frac{(x_1x_2)^{r/2}}{\Gamma(r+1)}\exp\left\{-\frac{\alpha(x_1+x_2)}{\lambda(1-\tau^2)}\right\} I_r\left\{\frac{2\alpha\tau\sqrt{(x_1x_2)}}{\lambda(1-\tau^2)}\right\},$$

where $\tau = \exp(-\frac{1}{2}\alpha t)$.

The right-hand side of equation (7.9) is the probability density function of a bivariate Gamma distribution with Laplace-Stieltjes transform

$$\phi(\theta_1,\theta_2) = \{1 + \frac{\lambda}{\alpha}\theta_1 + \frac{\lambda}{\alpha}\theta_2 + (1-e^{-\alpha t})\frac{\lambda^2}{\alpha^2}\theta_1\theta_2\}^{-(\kappa-\nu)/\lambda} \quad (7.10)$$

(see Vere-Jones (1967)). More generally, using equation (7.7) together with the Markov property of the process, it may be shown that the joint distribution of $Z(t_1),Z(t_2),\ldots,Z(t_n)$ is a special case of a multivariate Gamma distribution considered by Krishnamoorthy and Parthasarathy (1951). The n-variate analogue to equation (7.10) is

$$E[e^{-\underset{\sim}{\theta}'\underset{\sim}{Z}(t)}] = \{\det(\underset{\sim}{I}+\underset{\sim}{A})\}^{-(\kappa-\nu)/\lambda}, \tag{7.11}$$

where $\underset{\sim}{A}$ is the matrix $\{a_{ij} = \lambda\alpha^{-1}\theta_j \exp(-\frac{1}{2}\alpha|t_i-t_j|)\}$.

In the case when $2(\kappa-\nu)/\lambda$ is a positive integer, p, say, Krishnamoorthy and Parthasarathy showed that Z(t) may be written as

$$Z(t) = \sum_{k=1}^{p} U_k^2(t),$$

where $\{U_k(t_i), k = 1,2,\ldots,p;\ i = 1,2,\ldots,n\}$ is a sample of k independent vectors from an n-variate Gaussian distribution with zero means and identical variances. In other words $\{U_k(t), k = 1,2,\ldots,p\}$ may be considered as a vector of k independent, identically distributed, zero-mean stationary Gaussian processes. Barndorff-Neilson and Yeo (1969) have investigated properties of the process Z(t) in the above representation. It may be noted that if the $U_k(t)$ are stationary Ornstein-Uhlenbeck processes, Z(t) has the covariance structure given by equation (7.11), and is thus identical to the process we have been considering. It follows that if $2(\kappa-\nu)/\lambda$ is a positive integer p this diffusion approximation to the linear growth birth, death and migration process may be regarded as the sum of squares of p independent stationary Ornstein-Uhlenbeck processes. It is interesting to note that the sum of squares of a number of independent identically distributed Markov Gaussian processes is also Markovian.

(b) Transient processes: For some of the models considered previously, it was possible to obtain convergence of the transformed sequence $\underset{\sim}{Z}_N(t) = \{\underset{\sim}{\alpha}_N(t) - \underset{\sim}{a}N\}/\sqrt{N}$ only for certain critical values of the parameters. Thus for the problem of the two sexes with intermediate dominance, we required $\lambda = \sqrt{(\mu_1\mu_2)}$. Convergence may be obtained under more general conditions if one allows a to depend on t, so that we generalize (4.2) to

$$\underset{\sim}{Z}_N(t) = \{\underset{\sim}{X}_N(t) - N\underset{\sim}{a}(t)\}/\sqrt{N}. \tag{7.13}$$

As an illustration of this type of situation we may consider the general stochastic epidemic, for which asymptotic results have already been obtained by Nagaev and Startsev (1970). This process is a Markov population (bivariate) process in which $X_1(t)$ is the number of infectives and $X_2(t)$ the number of susceptibles. The birth, death and transference rates are defined as

$$\begin{cases} \lambda_1(\underset{\sim}{n}) = \lambda_2(\underset{\sim}{n}) = 0 \\ \mu_1(\underset{\sim}{n}) = \gamma n_1, \ \mu_2(\underset{\sim}{n}) = 0 \\ \gamma_{12}(\underset{\sim}{n}) = 0, \ \gamma_{21}(\underset{\sim}{n}) = \beta n_1 n_2 \end{cases} \tag{7.14}$$

Replacing β by β/N, and considering the transitions in $(t,t+\delta t)$, we find

$$E[e^{i\underset{\sim}{\theta}'\underset{\sim}{Z}_N(t+\delta t)}]e^{i\sqrt{N}\underset{\sim}{\theta}'\{\underset{\sim}{a}(t+\delta t)-\underset{\sim}{a}(t)\}} - E[e^{i\underset{\sim}{\theta}'\underset{\sim}{Z}_N(t)}]$$
$$= \beta\{e^{i\theta_1/\sqrt{N}} + e^{-i\theta_2/\sqrt{N}} - 2\}E[X_1(t)X_2(t)e^{i\underset{\sim}{\theta}'\underset{\sim}{Z}_N(t)}]\delta t \tag{7.15}$$
$$+ \gamma\{e^{i\theta_1/\sqrt{N}} - 1\}E[X_1(t)e^{i\underset{\sim}{\theta}'\underset{\sim}{Z}_N(t)}]\delta t + o(\delta t).$$

Taking the limit as $\delta t \to 0$, and using equation (7.13), we derive

$$\frac{\partial}{\partial t}\phi_N + i\sqrt{N}\underset{\sim}{\theta}'\frac{\partial}{\partial t}\underset{\sim}{a}(t)\phi_N$$
$$= N^{-1}\beta\{e^{i\theta_1/\sqrt{N}} + e^{-i\theta_2/\sqrt{N}} - 2\}\{N^2 a_1(t)a_2(t)$$
$$- iN\sqrt{N}[a_2(t)\frac{\partial}{\partial\theta_1} + a_1(t)\frac{\partial}{\partial\theta_2}]\}\phi_N$$
$$+ \gamma\{e^{-i\theta_1/\sqrt{N}} - 1\}\{Na_1(t) - i\sqrt{N}\frac{\partial}{\partial\theta_1}\}\phi_N + 0(1/\sqrt{N}), \tag{7.16}$$

where $\phi_N = E[\exp\{i\underset{\sim}{\theta}'\underset{\sim}{Z}_N(t)\}]$. If ϕ_N is to approach a finite limit as $N \to \infty$, it is necessary that the $0(\sqrt{N})$ terms in equation (7.16) cancel. Thus

$$\begin{cases} \dot{a}_1(t) = \beta a_1(t)a_2(t) - \gamma a_1(t) \\ \dot{a}_2(t) = -\beta a_1(t)a_2(t), \end{cases} \tag{7.17}$$

where the dot represents differentiation with respect to t. Now taking the limit as $N \to \infty$ in equation (7.16), we expect

$$\begin{aligned} \frac{\partial\phi}{\partial t} &= -\tfrac{1}{2}\{\beta(\theta_1^2+\theta_2^2)a_2(t) + \gamma\theta_1^2\}a_1(t)\phi \\ &\quad + \theta_1\{\beta a_2(t)-\gamma\}\frac{\partial\phi}{\partial\theta_1} + \theta_2\beta a_1(t)\frac{\partial\phi}{\partial\theta_2} \end{aligned} \tag{7.18}$$

The major difference between equation (7.18) and previous analogues such as (4.8) is that the coefficients are now time-dependent. However it is still possible to obtain explicit solutions; in the absence of boundary conditions the diffusion characterized by equation (7.18) is Gaussian, and its (time-dependent) mean and covariance nature are obtainable by substitution. Equations (7.17) will be recognized as the equations characterizing the "deterministic model". Thus the general stochastic epidemic, when transformed according to equation (7.13), converges in probability to the deterministic curve given by the solution to equations (7.17), in such a way that the residual (multiplied by $\sqrt{N}$) is a bivariate Gaussian diffusion process whose characteristic function satisfies the forward Kolmogorov equation (7.18).

The above example clarifies the relationship between the deterministic models which have been suggested by biologists and other natural and social scientists and the stochastic models formulated by the statisticians. There are also parallels in econometrics, where there is a need to relate linear models with additive error terms and stochastic models. The above approach may point the way to a better understanding of the interplay between these models.

(c) Queues with time-dependent arrival rates: A number of limit laws which are special cases of the more general theorem discussed in section 2 were obtained by J.F.C. Kingman (see, for

example, Kingman (1964)). These results apply to queues in "heavy traffic". In fact Kingman showed that the same limiting results apply to a wide class of queueing situations, not just those characterized by birth and death processes. Our object here is to suggest that Kingman's results may be generalized to queueing situations in which (a) the arrival rates are time-dependent and (b) the service rate depends in a general way upon the content of the system.

Let $X(t)$ denote the content at time t of a Markovian queue with non-homogeneous simple Poisson arrivals having intensity $\lambda(t)$ and departures which constitute a pure death process. Thus $X(t)$ is a "non-homogeneous" birth and death process with birth rate $\lambda(n) = \lambda(n,t) = \lambda(t)$ and death rate $\mu(n)$. We now consider a sequence of such processes, $\{X_N(t)\}$, with rates $\lambda_N(t)$ and $\mu_N(n)$, and define

$$Z_N(t) = \{X_N(t) - Na(t)\}/\sqrt{N}, \tag{7.19}$$

where $a(t)$ is a function to be determined. Provided

$$\begin{aligned} \lambda_N(t) &= N\alpha(t), \\ \mu(Na(t) + x\sqrt{N}) &= N\xi(a(t)) + x\sqrt{N}\xi'(a(t)) + 0(1), \end{aligned} \tag{7.20}$$

the heuristic method of section 2 yields

$$\frac{\partial\phi}{\partial t} = -\frac{1}{2}\theta^2\{\alpha(t) + \xi(a(t))\}\phi - \theta\xi'(a(t))\frac{\partial\phi}{\partial\theta}, \tag{7.21}$$

where $\phi = \phi(\theta,t) = \lim_{N\to\infty} E[\exp\{i\theta Z_N(t)\}]$, and $a(t)$ must satisfy the equation

$$\dot{a}(t) = \alpha(t) - \xi(a(t)). \tag{7.22}$$

In the special case when the service rate does not depend on the content of the system, it is appropriate to put $\mu_N(n) = N\mu$, so that $\xi(a(t)) \equiv \mu$, $\xi'(a(t)) \equiv 0$. Thus equation (7.22) yields

$$a(t) = a(0) + \int_0^t \{\alpha(n) - \mu\}du. \tag{7.23}$$

In particular, if the arrival rate is constant and equal to μ, $a(t)$ is constant, and equation (7.21) reduces to

$$\frac{\partial\phi}{\partial t} = -\mu\theta^2\phi, \tag{7.24}$$

which characterizes a Wiener process with variance $2\mu t$. A subtler limit results if one puts $\mu_N(n) = \mu N + \beta\sqrt{N}$. In this case equation (7.20) does not hold, but the analogue to (7.21) is

$$\frac{\partial\phi}{\partial t} = -\frac{1}{2}\theta^2\{\alpha(t) + \mu\}\phi - \theta\beta\frac{\partial\phi}{\partial\theta}. \tag{7.25}$$

In the case $\alpha(t) \equiv \mu$, this reduces to the equation for a Wiener process with downward or upward drift, according as $\beta > 0$ or $\beta < 0$. The deterministic part of the motion is, from (7.23), $a(t) = \text{const}$. This is precisely the asymptotic result obtained by Kingman.

It should be mentioned that Newell (1968) has investigated diffusion approximations to queues with time-dependent arrival rates, but his method is based on equating means and variances of the process and a non-stationary Wiener process with drift.

REFERENCES

Bailey, N. T. J. (1964). The Elements of Stochastic Processes with Applications to the National Sciences. Wiley, New York.

Barndorff-Nielson, O. and Yeo, G. F. (1969). Negative binomial process, *J. Appl. Prob.*, *6*, pp. 633-647.

Bellman, R. (1970). Introduction to Matrix Analysis, (2nd edition). McGraw-Hill, New York.

Chandrasekhar, S. (1943). Stochastic problems in physics and astronomy, *Rev. Mod. Phys.*, *15*, pp. 1-89.

Cox, D. R. and Miller, H. D. (1965). The Theory of Stochastic Processes. Wiley, New York.

Feller, W. (1968). An Introduction to Probability Theory and Its Applications, Vol. II (2nd ed). Wiley, New York.

Goodman, L. A. (1953). Population growth of the sexes, Biometrics, *9*, pp. 212-225.

Iglehart, D. L. (1965). Limiting diffusion approximations for the many server queue and the repairman problem, *J. Appl. Prob.*, *2*, pp. 429-441.

Kendall, D. G. (1949). Stochastic process and population growth, *J. Roy. Statist. Soc. (B)*, *11*, pp. 230-264.

Kingman, J. F. C. (1964). The heavy traffic approximation in the theory of queues, paper in "Congestion Theory," (W. L. Smith and W. E. Wilkinson, editors), University of N. Carolina Press, Chapel Hill (1965).

Kingman, J. F. C. (1969). Markov population processes, *J. Appl. Prob.*, *6*, pp. 1-18.

Krishnamoorthy, A. S. and Parthasarathy, M. (1951). A Multivariate Gamma-type distribution, *Annals of Math. Statist.*, *22*, pp. 549-557.

Lampard, D. G. (1968). A stochastic process whose successive intervals between events form a first order Markov chain - I, *J. Appl. Prob.*, *5*, pp. 648-668.

Nagaev, A. V. and Startsev, A. N. (1970). The asymptotic analysis of a stochastic model of an epidemic, *Theory of Prob. Applicns.*, *15*, pp. 98-107.

Newell, G. F. (1968). Queues with time-dependent arrival rates I, II and III, *J. Appl. Prob.*, *5*, pp. 436-451, 579-590, 591-606.

Prendiville, B. J. (1949). Discussion of paper by Kendall, *J. Roy. Statist. Soc. (B)*, *11*, p. 273.

Schach, S. (1970). Markov models for multi-lane freeway traffic, Transportation Research, *4*, pp. 259-266.

Schach, S. (1971). Weak convergence results for a class of multivariate Markov processes, *Ann. Math. Statist.*, *42*, pp. 451-465.

Vere-Jones, D. (1967). The infinite divisability of a bivariate Gamma distribution, Sankya, (A), *29*, pp. 421-422.

Whittle, P. (1967). Nonlinear migration processes, Bulletin of the International Statistical Institute, *42*, pp. 642-647.

THE STOCHASTIC NATURE OF PENSION COSTS

Arnold F. Shapiro

Department of Business Administration
The Pennsylvania State University
University Park, Pennsylvania

SUMMARY

Given that expected pension costs are used as a basis for funding a pension plan, two questions immediately arise. First, what is the probability that the expected cost will be at least as great as the actual cost? Second, given that the probability of this occurrance is not satisfactory, what additional funds are required to bring the probability to a satisfactory level? An approach to these two problems is discussed in this paper.

INTRODUCTION

Most studies of pension costs use a deterministic model, that is, a model which is based entirely on expected values. While it is true that the use of expected values results in a "best estimate" of expected cost, the probability that the expected cost will actually obtain may be very small. Furthermore, a pension

ISBN 0-12-394680-8

plan which is funded to the extent of its expected cost may or may not have sufficient funds to meet its actual cost. Thus, a deterministic model may have serious shortcomings when used as a guide in the funding of a pension plan.

The deficiencies of a deterministic model may be traced directly to the fact that many of the parameters which enter into the determination of expected pension costs are random variables, and, as such, they may take on values which are significantly different from their expected values. In order to overcome these deficiencies a stochastic, or probabilistic, model is needed.

Very few articles have dealt specifically with the development of a stochastic model of pension costs. Stone [12], in 1948, investigated the impact of mortality fluctuations on pensions paid to pensioners. The main thrust of that study was the use of probability generating functions to develop probabilities, at various durations after employees had begun to retire, that the sum total of actual pension payments would differ from the expected total payments. Taylor [13], in 1952, investigated the size of the contingency reserve needed to insure, with a given probability, that the funds on hand would be sufficient to pay all promised pensions. Both these studies dealt exclusively with the retired population, under the assumption that the number of retirees was known. One of the few studies to deal explicitly with the active population of a pension plan was a study by Seal [11], in 1953, in which he dealt with the problem of determining the mathematical risk of lump sum death benefits in a trusteed pension plan. Under the assumption of a "reasonably" large group of employees, Seal introduced variance minimization into the design of pension plans.

While the aforementioned studies were among the most notable in the pension area, per se, it should not be inferred that these are the only relevant studies. In fact, to the extent that pensions may be regarded as annuities, there have been a considerable number of relevant studies. Piper [9], in 1933, for example, developed contingency reserves for life annuities based on the mean and variance associated those annuities. Menge [8], in 1937, and Hickman [6], in 1964, elaborated on the Piper article: Menge using discrete functions and Hickman using continuous functions. Hickman's article, in addition, extended the development to include loss functions and a probablistic consideration of multiple decrement theory.

Although it is clear that the number of lives which persist to a given age from an initial group of lives is generated by a Bernoulli process, the considerable labor required to generate appropriate distributions under this process resulted in the development of various approximation methods. Hence, Piper assumed a large group of lives and used a normal distribution, as did Seal; Taylor suggested fitting a Pearson Type III distribution to the total present value of life annuity costs; Boermeester [1], in 1956, applied a Monte Carlo approach to the problem; Fretwell and Hickman [4], in 1964, investigated upper bounds for the cost using the inequalities of Tchebychef and Uspensky; and Bowers [2], in 1968, investigated the use of the Cornish-Fisher expansion to develop probabilities of sufficient reserves, based on correction factors applied to a standard normal table.

The capabilities of computers has been improved considerably since these studies were done. In this study it is shown that it may now be feasible to make probability statements regarding pension costs without resorting to limiting distributions.

THE PROBABILITY OF A GIVEN NUMBER OF EMPLOYEES AT EACH AGE

The expected number of employees at age x, $\bar{\ell}_x^{aa}$, is usually derived using the recursion formula

$$\begin{aligned} \bar{\ell}_x^{aa} &= \ell_a^{aa}, && x = a \\ &= \bar{\ell}_{x-1}^{aa} \cdot p_{x-1}^{aa}, && x > a \end{aligned} \qquad (1)$$

where p_{x-1}^{aa} is the expected probability of persisting through age x - 1, and "a" is the entry age. While $\overline{\ell}_x^{aa}$ represents the mathematical expectation or average number of employees at age x, the probability that this precise number will actually obtain may not be very high. This section quantifies the probability of a given number of employees at a given age.

Total Attribute Groups

To facilitate the development of the model it is convenient to segregate the pension population by qualification ages. This may be done in two steps. In the first step the types of qualification ages are isolated. Hence, all possible entry ages are grouped, all possible initial vesting ages are grouped, and so on. In the second step each type of qualification age is partitioned according to unique ages. For example, there may be ten unique entry ages, ten unique initial vesting ages, and so on.

The process may be quantified as follows. Let N_k denote the number of unique qualification ages with the attribute k. Similarly, let k^j, $j = 1,2,...,N_k$, denote the j-th unique qualification age with the attribute k. Finally, let C_k denote the set of unique qualification ages with the attribute k. Using this convention, the sets of unique qualification ages are given by

$$C_k = \{k^j: j = 1,2,...,N_k\} \tag{2}$$

Define a "total attribute" group as a group having exactly one attribute from each of the sets of unique qualification ages, and let "C" denote the set of all total attribute groups.

It should be mentioned that the "uniqueness" of the qualification ages is to be interpreted in a computational sense. Hence, for example, two employees whose entry age nearest birthday is twenty might both be given an entry age of twenty for computational purposes, even though they were not, in fact, the exact same age. Another computational convenience that is often used is to classify entry ages into quinquennial age groupings. Under this procedure each grouping would constitute a unique entry age.

An example of a total attribute group would be those active participants with an entry age of 20, an initial vesting age and initial disability qualification age of 30, an early retirement age of 50, a normal retirement age of 60, and a mandatory retirement age of 65. This particular total attribute group would be denoted by the six-tuple (20,30,30,50,60,65).

The concept of a total attribute group has seen generous implementation in the pension literature, albeit in a somewhat disguised form. Many articles, for example, have used "select" groups, where the common attribute has been the entry age. However, in most of the articles the other qualification ages were the same for each member of the select group. This meant that each select group had all qualification ages in common and was, in fact, a total attribute group.

A General Statement of the Probability of a Given Number of Employees at a Given age

In general, the number of employees at age x, from the total attribute group c, $c \in C$, may be regarded as a random variable, ${}^{c}\tilde{\ell}_{x}^{aa}$ say, that depends on the number of employees at the previous age, ${}^{c}\tilde{\ell}_{x-1}^{aa}$, which is also a random variable. Thus, the probability that ${}^{c}\ell_{x}^{aa}$ takes on some particular value, N say, that is, $\Pr\{{}^{c}\ell_{x}^{aa} = N\}$, must be determined by a recursive procedure. Let

$$ {}^{k}\underset{\sim}{\ell}^{aa} = ({}^{k}\ell_{x-t}^{aa} : t = 0,1,\ldots,x-a)' \tag{3}$$

denote a vector of ℓ_{x-t}^{aa} values consistent with a final value of $\ell_{x}^{aa} = N$, and call this vector a feasible ℓ_{x}^{aa} array. Assuming that there are K distinct feasible ℓ_{x}^{aa} arrays, it follows that

$$\Pr\{\ell_x^{aa} = N\} = \sum_{k=1}^{K} \Pr\{{}^k\ell_x^{aa}\}. \tag{4}$$

Denoting by

$$f(\ell_{x-t}^{aa} | \ell_x^{aa}, p_{x-t-1}^{aa}, \ell_{x-t-1}^{aa}) \tag{5}$$

the probability that exactly ℓ_{x-t}^{aa} employees will persist through age x-t-1, consistent with ℓ_x^{aa} employees persisting through age x, it follows that

$$\Pr\{{}^k\underset{\sim}{\ell}_x^{aa}\} = \prod_{t=0}^{x-a-1} f({}^k\ell_{x-t}^{aa} | \ell_x^{aa}, p_{x-t-1}^{aa}, {}^k\ell_{x-t-1}^{aa}). \tag{6}$$

From this formulation it is clear that the probability that exactly ℓ_x^{aa} employees persist to age x is

$$\sum_{k=1}^{K} \prod_{t=0}^{x-a-1} f({}^k\ell_{x-t}^{aa} | \ell_x^{aa}, p_{x-t-1}^{aa}, {}^k\ell_{x-t-1}^{aa}). \tag{7}$$

A Conditional Probability Distribution Function for ℓ_x^{aa}

In order to implement equation 7 it is necessary to specify the probability distribution function given by equation 5. Fortunately, under the assumption that valuations are based only on curtate ages, the number of employees who persist through a given age may be thought of as being generated by a Bernoulli process under which employees either persist as active members of leave the active group [10: Chap. 9]. It follows immediately that the conditional distribution of ℓ_{x-t}^{aa} given p_{x-t-1}^{aa} is specified by the binomial mass function [10:213]

$$f_b(\ell_{x-t}^{aa} | \ell_x^{aa}, p_{x-t-1}^{aa}, \ell_{x-t-1}^{aa})$$

$$= \binom{\ell_{x-t-1}^{aa}}{\ell_{x-t}^{aa}} (p_{x-t-1}^{aa})^{\ell_{x-t}^{aa}} (1 - p_{x-t-1}^{aa})^{\ell_{x-t-1}^{aa} - \ell_{x-t}^{aa}} \tag{8}$$

and the probability that exactly ℓ_x^{aa} employees persist to age x is

$$\Pr\{\ell_x^{aa}\} = \sum_{k=1}^{K} \prod_{t=0}^{x-a-1} f_b({}^k\ell_{x-t}^{aa} \mid \ell_x^{aa},\ p_{x-t-1}^{aa},\ {}^k\ell_{x-t-1}^{aa}). \tag{9}$$

It is important to recognize that the binomial mass function is appropriate only under the assumption that the exact probability of persisting is known. This assumption, however, may be inappropriate. Although it is true that estimates of p_x^{aa} are often available, these estimates may or may not be valid for the particular pension plan under consideration. Nonetheless, for the purpose of the present study, it is assumed that available data is representative of the pension population under consideration.[1]

THE PROBABILITY THAT EXPECTED RETIREMENT COSTS EXCEED ACTUAL RETIREMENT COSTS

The retirement cost associated with any particular total attribute group is

$${}^c\tilde{\ell}_r^{aa} \cdot {}^{cB}\bar{a}_r^{rr}, \quad c \in C, \tag{10}$$

where ${}^{cB}\bar{a}_r^{rr}$ represents the present value, at the age of retirement, age r, of the pension benefits. ${}^{cB}\bar{a}_r^{rr}$ is assumed to be given. The probability that the expected retirement cost for this group exceeds the actual pension cost is equal to

$$\Pr\{{}^c\bar{\ell}_r^{aa} \cdot {}^{cB}\bar{a}_r^{rr} \geq {}^c\tilde{\ell}_r^{aa} \cdot {}^{cB}\bar{a}_r^{rr}\}, \tag{11}$$

[1] *The development of distributions for the probabilities of decrement when the appropriateness of the data is questionable and an investigation of their impact on the adequacy of expected pension costs is currently being undertaken by the author, and will form the basis of a sequel to the present study.*

which reduces to[2]

$$\Pr\{{}^{c}\bar{\ell}^{aa}_{r} \geq {}^{c}\tilde{\ell}^{aa}_{r}\}. \tag{12}$$

In view of the integral properties of the function ${}^{c}\ell^{aa}_{r}$, this latter probability becomes[3]

$$\sum_{{}^{c}\ell^{aa}_{r}=0}^{[{}^{c}\bar{\ell}^{aa}_{r}]} \Pr\{{}^{c}\ell^{aa}_{r}\}. \tag{13}$$

Most pension plans, of course, have entrants at more than one age, so it is appropriate to extend the foregoing analysis to recognize this situation.

In general, the probability that the total expected retirement cost exceeds the total actual retirement cost is

$$\Pr\{\sum_{c} {}^{c}\bar{\ell}^{aa}_{r} \cdot {}^{cB}\bar{a}^{rr}_{r} \geq \sum_{c} {}^{c}\tilde{\ell}^{aa}_{r} \cdot {}^{cB}\bar{a}^{rr}_{r}\}. \tag{14}$$

The solution to equation 14 is facilitated by defining two arrays: a retirement benefit array and a feasible retirement array. Let

$$\underset{\sim}{{}^{B}\bar{a}^{rr}_{r}} = ({}^{cB}\bar{a}^{rr}_{r} \mid c \in C)' \tag{15}$$

be defined as the retirement benefit array associated with the pension plan under consideration, that is, the array whose elements are the present value of the retirement benefits, at retirement, associated with each total attribute group. Additionally, let

[2] *Note that for a particular total attribute group, the probability that the expected pension cost will be adequate is independent of the benefit function defined by the plan, assuming that ${}^{cB}\bar{a}^{rr}_{r}$ is given.*

[3] *The function* $[m]$ *represents the largest integer in* m.

$$ {}^{n}\ell^{aa}_{\tilde{r}} = ({}^{nc}\ell^{aa}_{r} \mid c \in C)' \tag{16}$$

be defined as a feasible retirement array, that is, an array whose elements are composed of possible numbers of employees who reach their normal retirement age from each total attribute group, and which satisfy the condition

$$ ({}^{n}\ell^{aa}_{\tilde{r}})^{T}\, B^{\overline{rr}}_{a_{\tilde{r}}} \leq \sum_{c} {}^{c}\bar{\ell}^{aa}_{r} \cdot {}^{c}B^{\overline{rr}}_{a_{r}} . \tag{17}$$

Assuming that there are N distinct feasible retirement arrays, it follows that the probability that the total expected retirement costs will exceed the total actual retirement cost is

$$ \sum_{n=1}^{N} \Pr\{{}^{n}\ell^{aa}_{\tilde{r}}\}. \tag{18}$$

However, since the probability of a given feasible retirement array is simply the product of the probabilities of the joint occurrance of each element of the array, the probability that the total expected cost will exceed the total actual cost becomes

$$ \sum_{n=1}^{N} \prod_{c} \Pr\{{}^{nc}\ell^{aa}_{r}\}. \tag{19}$$

The Contingency Charge

The determination of the contingency charge needed to increase the probability of adequate funds to a given level follows immediately from the foregoing analysis. The only change is that, instead of defining a feasible array in terms of expected cost, one would define a contingent feasible array in terms of some multiple of the expected cost. For example, one might define a contingent feasible retirement array as a feasible retirement array which satisfies the condition

$$ ({}^{n}\ell^{aa}_{\tilde{r}})^{T}\, B^{\overline{rr}}_{a_{\tilde{r}}} \leq (1+m)\sum_{c} {}^{c}\bar{\ell}^{aa}_{r} \cdot {}^{c}B^{\overline{rr}}_{a_{r}} , \tag{20}$$

where the factor (1+m) defines the multiple of the expected cost which is to be funded, and where the product of m and the expected cost represents the contingency charge.

In practice the factor (1+m) would be determined such that the probability of adequare funds obtains some desirable level, ninety-nine percent, for example.

WORKING FORMULAE FOR THE DETERMINATION OF THE PROBABLE ADEQUACY OF EXPECTED PENSION COSTS

Given that the number of participants at a given age is binomially distributed, it is a simple matter to set down a working formula for the probable adequacy of the expected pension cost. For a specific total attribute group, the probability that the expected retirement cost exceeds the actual retirement cost is

$$\sum_{c_{\ell_r^{aa}}=0}^{[c\overline{\ell}_r^{aa}]} \sum_{k=1}^{K} \prod_{t=0}^{x-a-1} f_b({}^{k}\ell_{r-t}^{aa} \mid {}^{c}\ell_r^{aa},\ p_{r-t-1}^{aa},\ {}^{k}\ell_{r-t-1}^{aa}). \qquad (21)$$

On the other hand, the probability that the total expected retirement cost exceeds the total actual retirement cost is

$$\sum_{n=1}^{N} \prod_{c} \sum_{k=1}^{K(c)} \prod_{t=0}^{{}^{c}r-{}^{c}a-1} f_b({}^{k}\ell_{r-t}^{aa} \mid {}^{nc}\ell_r^{aa},\ p_{r-t-1}^{aa},\ {}^{k}\ell_{r-t-1}^{aa}). \qquad (22)$$

A working formula for the probable adequacy of some multiple of the expected cost is similarly defined.

The final section of this study deals with the application of these formulae.

ESTIMATES OF THE ADEQUACY OF EXPECTED PENSION COSTS

Turning now to examples of estimates of the adequacy of expected pension costs, consider first the probability that the expected cost will exceed the actual cost, given a specific total

TABLE 1

The Proportion by Which the Expected Cost Must be Multiplied To Insure, Using a Ninty-nine Percent Confidence Level, That a Comparable Fund Will be Sufficient to Cover the Retirement Benefits, Given Various Numbers of Entrants at Age Twenty

Entrants At Age Twenty	Decrement Data										
	GAM71	T-01	T-02	T-03	T-04	T-05	T-06	T-07	T-08	T-09	T-10
50	1.14	1.62	1.66	1.69	2.03	2.14	3.60	5.20	4.38	6.09	2.29
100	1.12	1.45	1.44	1.45	1.72	1.78	3.00	3.47	3.65	4.06	1.86
150	1.09	1.37	1.37	1.38	1.62	1.66	2.80	2.89	2.92	3.38	1.72
200	1.08	1.30	1.31	1.34	1.52	1.60	2.40	2.60	2.56	3.05	1.65

attribute group. Table 1 shows the multiple by which the expected cost must be multiplied to insure, using a ninty-nine percent confidence level, that a comparable fund will be sufficient to cover retirement benefits, assuming that entry takes place at age twenty and that retirement takes place at age sixty-five. The decrements from the active population are mortality, based on the GAM 1971 Mortality Table [5]; disablement, based upon values tabulated by Winklevoss [15:40]; and withdrawal, based upon ten turnover tables given by McGinn (T-01, T-02,...,T-10) [7:235-6]. The latter are shown in Figure 1. To provide a standard of comparison, the first column of values was calculated under the assumption that there were no decrements other than mortality prior to age sixty-five.

Certain observations are immediately apparent from the table. First, the higher the probability of withdrawal during the earlier years the greater must be the multiple by which the expected cost must be multiplied in order to obtain a given likehood of solvency. Using Table I (T-01), for example, the required multiple, given two hundred entrants, annually, at age twenty, is only one-half of the required multiple using Turnover Table VII (T-07). Second, the multiple needed decreases asymptotically as the number of entrants increases. The average multiple of the expected cost needed to insure adequate funds, at the ninty-nine percent confidence level, decreases from a high of 3.07 for fifty entrants at age twenty to 1.93 for two hundred entrants. With respect to this latter observation, it is important to emphasize that two hundred entrants at the single entry age of twenty might result in a significantly large population. Under Turnover Table I, for example, two hundred entrants at age twenty, annually, would generate a mature population of 4,356 active employees. Thus, a plan of considerable size may be required to reduce significantly the contingency charge needed to attain a given confidence level.

Consider now the determination of the contingency charge when

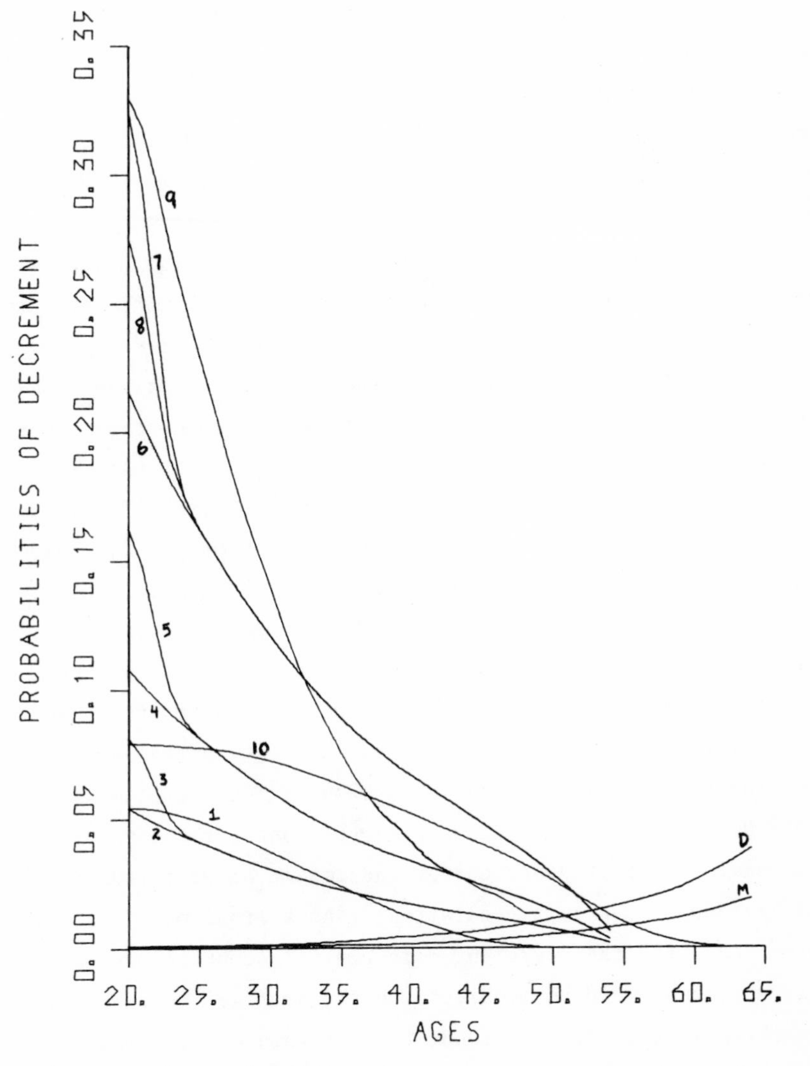

Fig. 1. The probability of turnover for entrants aged twenty.

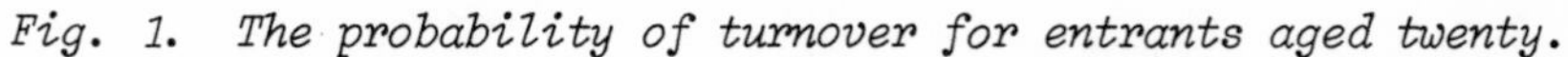

Source: McGinn's Turnover Tables

multiple entry ages are used. For the purpose of example, it is assumed that entry takes place quinquennially from ages twenty through fifty, inclusive, with the proportion of total entrants at each entry age being .28, .24, .18, .12, .08, .05, and .05, respectively [14:34]. The total number of entrants is chosen so that, if entry were to take place annually, a mature population of two thousand employees would result. In addition, the benefit function is assumed to be two percent of final salary for each year of service, using Salary Scale 3 of the "Actuary's Pension Handbook" [3].

Table 2 gives the size of the contingency charge needed to increase the probability of adequate funds to ninty-nine percent, given a lower entry age of twenty and various upper entry ages. It is clear that increasing the number of entry ages has somewhat the same impact as increasing the size of the population, in the sense that the larger the number of entry ages the smaller the contingency charge needed to obtain a given probability of adequate funds. Once again, the impact is asymptotic. It would appear that, other things being equal, the inclusion of additional entry ages, over and above some optimal number, might have only negligible incremental impact on the size of the contingency charge needed. It is interesting to note, also, that the impact of selection varies inversely as the number of entry ages. This is evident upon comparing Tables T-02 with T-03, T-04 with T-05, and T-06 with T-07 and T-08. It may be that for the purpose of determining a contingency charge the impact of selection may be disregarded for certain populations. As a final observation, one notes that, for the decrement data given, a contingency charge of forty percent would be required, on the average, to reduce the probability of inadequate funds to less than one percent.

TABLE 2

The Proportion by Which the Expected Cost Must be Multiplied to Insure, Using a Ninty-nine Percent Confidence Level, That a Comparable Fund Will be Sufficient to Cover the Retirement Benefits, Given Two Thousand Participants and a Lowest Entry Age of Twenty

Highest Entry Age	Decrement Data										
	GAM71	T-01	T-02	T-03	T-04	T-05	T-06	T-07	T-08	T-09	T-10
20	1.24	1.81	1.84	1.84	2.25	2.37	4.07	4.75	4.36	5.25	2.44
25	1.19	1.58	1.60	1.60	1.80	1.83	2.36	2.59	2.51	2.41	1.97
30	1.17	1.46	1.48	1.49	1.60	1.63	1.89	1.96	1.95	1.78	1.74
35	1.16	1.41	1.43	1.43	1.51	1.53	1.69	1.71	1.71	1.54	1.61
40	1.15	1.38	1.40	1.40	1.46	1.47	1.56	1.57	1.57	1.45	1.54
45	1.14	1.36	1.38	1.38	1.43	1.44	1.50	1.50	1.50	1.41	1.50
50	1.14	1.35	1.37	1.37	1.41	1.41	1.44	1.44	1.44	1.37	1.47

Conclusion

Although the examples used in this study are far from exhaustive, there appears to be considerable evidence that expected cost estimates for a given generation of retirees might significantly understate actual costs. The probability of this occurance depends, of course, on the particular pension plan and population being considered, but the appropriateness of funding only to the extent of expected pension costs may be questionable.

COMMENT

While it is clear that a stochastic study of pension cost is superior to a deterministic study, in the sense that it introduces credibility into the pension cost estimates, there have been very few stochastic studies of pension costs. This fact leads one to enquire into the reason for this deficiency in the pension literature. It is clear that the technology is available, the present paper shows that such a study is possible. Hence, it is something other than the lack of technology that has caused the lack of stochastic pension cost models. Two obvious possibilities are lack of interest and lack of resources. The first of these possibilities is easily expelled. For many years actuaries have been concerned with the limitations imposed by expected cost estimates. Rather than use stochastic models, however, many investigators have choosen, instead, to provide a number of pension cost estimates based upon a spectrum of assumptions. Whether these researchers would have used a stochastic model, if it were available, is a matter of conjecture. However, there appears to be no question that the limitations imposed by a deterministic model is an area of significant concern, and it seems reasonable to rule out a lack of interest as a reason for not employing stochastic models.

A more feasible reason for not employing stochastic models is a lack of resources. In this respect the considerable computer memory and execution time need for implementation are likely to be the most critical limitations. In the first place, the study should be done in double precision, since rounding errors tend to accumulate very quickly. In addition, depending on the scope of the model--whether it can accommodate multiple entry ages, the cost of vesting, *et cetera* -- the size of the total active population which can be investigated may be limited.

In most instances the amount of time needed to execute the model is probably its most severe limitation, since execution time increases geometrically as the size of the population and/or number of entry ages increases. The determination of an appropriate contingency charge when multiple retirement ages are used or when the cost of vesting is considered would also require considerable execution time.

It is important to mention that the foregoing limitations of the model may be a function of the program which was used. While it is true that every effort was made to reduce execution time and that the most efficient form of fortran was used, Fortran H (OPT = 2), the program may not be as efficient as it could be, and, indeed, the development of a "most efficient" model is clearly an area for future study.

The above observations notwithstanding, it is hoped that the model which was developed in this paper will help stimulate both theoretical and empirical research into the stochastic nature of pension costs. As regards the former, there are many refinements which might be incorporated into the model, including such things as a stochastic accumulation of funds, an unconditional distribution for the probability of persisting, and a stochastic retirement annuity. As regards the latter, although this paper did attempt to obtain certain specific results, these results were intended primarily as examples of the implementation of the model, and, as such, were far from exhaustive. Future researchers should find the empirical study of the stochastic nature of pension costs a fruitful area for exploration, particularly if they have at their disposal considerable computer facilities.

REFERENCES

[1] Boermeester, J. M. "Frequency Distribution of Mortality Costs." Transactions of the Society of Actuaries, Vol. VIII (1956), pp. 1-9.

[2] Bowers, N. L. "An Approximation to the Distribution of Annuity Costs," Transactions of the Society of Actuaries, Vol. XIX (1968), pp. 295-309.

[3] Crocker, Thomas F., Sarason, Harry M., and Straight, Byron W. The Actuary's Pension Handbook, Los Angeles, Cal.: Pension Publications, 1973.

[4] Fretwell, Robert L., and Hickman, James C. "Approximate Probability Statements About Life Annuity Costs," Transactions of the Society of Actuaries, Vol. XVI (1964), pp. 55-60.

[5] Greenlee, Harold R., and Keh, Alfonso D. "The 1971 Group Annuity Mortality Table," Transactions of the Society of Actuaries, Vol. XXIII (1971), pp. 569-604.

[6] Hickman, James, C. "A Statistical Approach to Premiums and Reserves in Multiple Decrement Theory," Transactions of the Society of Actuaries, Vol. XVI (1964), pp. 1-16.

[7] McGinn, Daniel F. "Indices to the Cost of Vested Pension Benefits," Transactions of the Society of Actuaries, Vol. XVIII (1966), pp. 187-242.

[8] Menge, W. O. "Statistical Treatment of Actuarial Functions," The Record of the American Institute of Actuaries, Vol. XXVI (1937), pp. 65-88.

[9] Piper, Kenneth B. "Contingency Reserves for Life Annuities," Transactions of the Actuarial Society of America, Vol. XXXIV (1933), pp. 240-249.

[10] Raiffa, Howard, and Schlaifer, Robert. Applied Statistical Decision Theory. Boston, Mass: Harvard Business School, 1961.

[11] Seal, Hilary L. "The Mathematical Risk of Lump-Sum Death Benefits in a Trusteed Pension Plan," Transactions of the Society of Actuaries, Vol. V (1953), pp. 135-142.

[12] Stone, David G. "Actuarial Note: Mortality Fluctuations in Small Self-Insured Pension Plans," Transactions of the Actuarial Society of America, Vol. XLIX (1948), pp. 82-91.

[13] Taylor, Robert H. "The Probability Distribution of Life Annuity Reserves and Its Application to a Pension System," Proceedings of the Conference of Actuaries in Public Practice, Vol. II (1952), pp. 100-150.
[14] Winklevoss Howard E. Analysis of the Cost of Vesting in Pension Plans. Washington, D.C.: U. S. Department of Labor, 1972.
[15] Winklevoss Howard E. Estimates of the Cost of Vesting in Pension Plans. Washington, D.C.: U. S. Government Printing Office, 1973.

AN APPROXIMATION METHOD TO CALCULATE THE VALUE OF A MATURITY GUARANTEE UNDER A LEVEL-PREMIUM-EQUITY-BASED CONTRACT

Phelim Paul Boyle

Faculty of Commerce and Business Administration
University of British Columbia
Vancouver, British Columbia

INTRODUCTION AND SIMULATION METHOD

The problem of valuing death benefit guarantees and maturity benefit guarantees under equity based insurance contracts has received considerable attention in recent years. The usual approach is to carry out a simulation study using actual stock market returns or statistics derived from these returns to estimate the value of the guarantee. In 1971 Kahn was able to find an analytical expression for the value of the death benefit guarantee under a particular type of single premium equity-based contract. The analytic approach permits one to gain a better appreciation of the investment risk involved.

In the present work the value of a maturity guarantee on a level premium equity based endowment policy is found by simulation and then using an approximation method. To concentrate on the investment risk we ignore mortality, lapses and expenses. Initially a simulation approach is used to value the guarantee. The nature

ISBN 0-12-394680-8

of the risk is explored by finding the first three moments of the value of the guarantee. By varying the assumed volatility of the underlying equity portfolio we can examine how this affects the value of the maturity guarantee. An approximation method, which does not use simulation, is then developed. We test the results of the method against those produced by simulation for portfolios of different volatilities. This method gives quite reasonable results for terms of up to 15 years. For longer terms it overstates the value of the guarantee.

The contracts analyzed consist of level premium n-year endowments where n runs from 1 to 30. It is assumed that premiums are paid annually. A fixed amount of each premium (the investment component) is invested in a portfolio of common stocks. On maturity it is guaranteed that the sum payable will not be less than the total amount invested. For the contracts considered we assume that the investment component of each premium is one unit.

The investment model assumes that the returns on the equity portfolio are lognormally distributed. This assumption has received considerable support in the financial literature. In addition Kahn used this model in the paper mentioned earlier. In particular we assume that the rate of return in year t is denoted by r_t. Our assumption is that $\log(1 + r_t)$ is normally distributed. We assume that the mean of $\log(1 + r_t)$ is μ and that its variance is σ^2. Notice that this implies that the returns in successive years are independent.

For our central case assumptions we take $\mu = 0.08$ and $\sigma = 0.2$. These values correspond to those that have been established by empirical studies except that standard deviation is on the high side since this increases the value of the guarantee and may make it easier to work with. Later on we will vary this parameter.

It is straightforward to generate a series of investment returns using a procedure for generating normally distributed variates. The results of one such set of 5,000 simulations are given in Table 1. Notice that for the type of contract and guarantee

TABLE 1. Maturity value under an n-year endowment policy assuming that the annual investment component is one unit and that the guaranteed amount is n units; $\mu = 0.08$, $\sigma = 0.2$.

Term	Expected value of guarantee	Standard deviation	Skewness coefficient
1	.042	.078	2.073
2	.073	.152	2.405
3	.100	.228	2.658
4	.122	.301	2.909
5	.142	.375	3.154
6	.158	.446	3.357
7	.174	.514	3.539
8	.185	.579	3.733
9	.193	.636	3.933
10	.199	.691	4.126
11	.204	.742	4.349
12	.207	.791	4.605
13	.206	.836	4.877
14	.206	.880	5.111
15	.205	.920	5.343
16	.205	.957	5.562
17	.204	.991	5.783
18	.200	1.015	6.028
19	.197	1.038	6.237
20	.192	1.054	6.471
21	.187	1.069	6.715
22	.179	1.075	6.973
23	.172	1.076	7.263
24	.164	1.074	7.639
25	.154	1.069	8.083
26	.147	1.073	8.425
27	.142	1.077	8.733
28	.138	1.085	8.996
29	.134	1.086	9.278
30	.130	1.084	9.602

considered the expected value of the guarantee at maturity increases to a maximum at around 12 years. (To find the percentage addition to the premium to cover the value of the guarantee these values would have to be discounted at an appropriate interest rate and respread to give a level annual equivalent. In addition allowance would have to be made for mortality and lapses).

The standard deviation of the value of the guarantee tends to be 5 to 6 times the expected value of the guarantee. The skewness of the distribution is an increasing function on n. The implications of the high skewness of the value of the guarantee have been pointed out by Kahn.

It is instructive to examine the sensitivity of the results with respect to charges in σ. As has been pointed out elsewhere this is a most critical varible in determining the value of the guarantee. A summary of the results is given in Table 2.

TABLE 2: Expected value of maturity guarantee under an n-year equity based endowment contract for $\mu = 0.08$ and various values of σ.

n	$\sigma = .10$	$\sigma = .20$	$\sigma = .30$
1	.012	.042	.074
5	.016	.142	.316
10	.007	.199	.535
15	.003	.205	.662
20	.000	.192	.709
25	.000	.154	.709
30	.000	.130	.667

THE APPROXIMATION METHOD

Let us assume that the total return on the equity portfolio at the end of n years is x. (For convenience we suppress the dependence on n). Suppose that x has the distribution function F(x). Then the value of the guarantee is equal to

$$\int_0^n (n-x)\,dF(x) \tag{1}$$

Integrals of this type are familiar to actuaries in connection with stop loss net premiums. In fact the formal expression for the stop loss net premium for losses in excess of Z is

$$\int_Z^{\infty} (x-Z)\,dF(x) \tag{2}$$

Just as in our case the precise form of the distribution function is usually unknown. However it is often possible to make progress if we know the moments of F(x). It turns out that in our problem the first few moments can be evaluated exactly and expressed in a compact form. These results are contained in the following theorem:

Theorem. Let $Y_1, Y_2, Y_3 \ldots Y_n$ be n independent random variables with

$$E(Y_i) = \rho$$

$$E[(Y_i)^2] = \theta \quad \text{for} \quad 1 \leq i \leq n$$

Let

$$x = Y_n + Y_nY_{n-1} + \ldots Y_nY_{n-1} \ldots Y_1. \tag{3}$$

Then

$$E(x) = F(n,\rho) \tag{4}$$

where

$$F(n,\rho) = \sum_{t=1}^{n} \rho^t,$$

and the variance of x is given by

$$F(n,\theta) \left[\frac{\theta+\rho}{\theta-\rho}\right] - \frac{2\theta}{[\theta-\rho]} F(n,\rho) - [F(n,\rho)]^2 \tag{5}$$

where

$$F(n,\theta) = \sum_{t=1}^{n} \theta^t.$$

The first result follows from the independent of the Y_i. The second result can be proved by induction. In the same way an expression for the third central moment can be found, but we do not give the details here. Notice that expressions (4) and (5) can be evaluated using a table of compound interest functions. The function x represents the total terminal return on the n-year investment program. The values of the mean and standard deviation for

TABLE 3: Terminal return on an investment of one per annum after n years assuming that equity returns are lognormally distributed:
$\mu = 0.08$, $\sigma = 0.20$

n	*Expected Value*	*Standard Deviation*
1	1.105	.223
2	2.327	.533
3	3.676	.955
4	5.168	1.500
5	6.817	2.181
6	8.639	3.016
7	10.653	4.024
8	12.878	5.230
9	15.338	6.661
10	18.056	8.350
15	36.587	22.104
20	67.138	50.460
25	117.509	106.676
30	200.558	215.552

various values of n are shown in Table 3.

In order to motivate the representation of x by a specific distribution function we note that x can be regarded as the outcome of n single premium contracts. Since the product of two lognormal variates is again a lognormal variate, x is the sum of n lognormal variates. Although the sum of lognormal variates *is not a lognormal variate* this can be a useful approximation in certain circumstances. For example in (2) Lintner states "We do claim on the basis of these simulations that the approximation to lognormally distributed portfolio outcomes of lognormally distributed stocks is sufficiently good that theoretical models based on these twin premises should be useful in a wide range of applications and empirical investigations." This leads us to represent x by a lognormal distribution.

From (5) we know that if x is lognormally distributed with mean M and standard deviation S then log x is normally distributed with mean MU and standard deviation SIG where

TABLE 4: Comparison of approximation method and simulation method for $\mu = 0.08$ and different values of σ.

Term	$\sigma = .10$		$\sigma = .20$		$\sigma = .30$	
	Approximation Method	Simulation	Approximation Method	Simulation	Approximation Method	Simulation
1	.011	.012	.042	.042	.071	.074
2	.015	.015	.073	.073	.135	.138
3	.017	.017	.102	.100	.200	.198
4	.017	.017	.129	.122	.266	.258
5	.016	.016	.153	.142	.332	.316
6	.015	.014	.174	.158	.398	.369
7	.014	.012	.194	.174	.465	.418
8	.012	.010	.211	.185	.532	.461
9	.011	.009	.227	.193	.599	.501
10	.010	.007	.242	.199	.667	.535
11	.008	.006	.255	.204	.734	.567
12	.007	.005	.267	.207	.802	.596
13	.006	.004	.278	.206	.870	.622
14	.006	.003	.287	.206	.937	.643
15	.005	.003	.296	.205	1.004	.662
20	.002	.000	.327	.192	1.326	.709
25	.001	.000	.340	.154	1.608	.709
30	.001	.000	.337	.130	1.831	.667

$$SIG = (\log(1 + \frac{S^2}{M^2}))^{\frac{1}{2}} \tag{6}$$

$$MU = \log M - \frac{(SIG)^2}{2} \; . \tag{7}$$

Thus for a given value of n we can use (6) and (7) to find the parameters of a lognormal distribution to represent x. This information enables us to obtain an explicit expression for the value of the guarantee using the method given in the Appendix to Kahn's 1971 paper. For an n year contract the value of the guarantee is

$$n\,N(\alpha) - M\,N(\beta) \tag{8}$$

where

$$\alpha = \frac{\log n - MU}{SIG}$$

$$\beta = \alpha - SIG \tag{9}$$

$$N(y) = \frac{1}{\sqrt{2\pi}} \int_{-\infty}^{y} e^{-w^2/2} dw.$$

In Table 4 the values of the naturity guarantee produced by this method are compared with these found earlier by simulation. The approximation is very good for $\sigma = .1$. For the other two values of σ the answers correspond well for terms up to 15 years. The approximation method in virtually all cases gives values higher than those produced by simulation. In view of the high skewness of the guarantee value, (given in Table 1) this may not be a completely undesirable feature. It may well be that this approach can be refined to give more accurate results.

ACKNOWLEDGEMENTS

The author has benefited by discussions with his colleagues Eduardo Schwarz and Michael Brennan with whom he is working on a different approach to this problem.

REFERENCES

[1] Kahn, P. M., "Projections of Variable Life Insurance Operations." Transactions of the Society of Actuaries, XXIII (1971), p. 335.

[2] Lintner, J., "Equilibrium in a Random Walk and Lognormal Securities Market." Discussion Paper No. 235, Harvard Institute of Economic Research, (1972) Harvard University, Cambridge, Massachusetts.

[3] Bowers, Newton, L., "An Upper Bound on the Stop-loss Net Premium-actuarial note." Transactions of the Society of Actuaries, XXI (1969), p. 211.

[4] Boyle, Phelim, P., "Rates of Return as Random Variables." Unpublished paper (1975). The University of British Columbia, Faculty of Commerce and Business Administration.

[5] Aitchison, J., and Brown, J.H.C., "The Lognormal Distribution." Cambridge University Press (1963).

EFFICIENT SORTING BY COMPUTER: AN INTRODUCTION

Robert Sedgewick[1]

Department of Computer Science
Brown University
Providence, Rhode Island

SUMMARY

This paper is a tutorial description of three important methods for sorting with a computer: heapsort, replacement selection, and polyphase merging. The first is appropriate when the file to be sorted fits entirely into the memory of the computer, and the others, in combination, are used for very large files.

The methods are given with step-by-step English language descriptions, and a full complement of examples is included. The paper is intended for readers who may have a little programming experience, but no familiarity with any particular computer or computer language is assumed.

It is hoped that the reader of this paper will not only learn some important sorting methods, but also gain some appreciation of a few basic concepts from the vast literature on sorting.

[1]*This work was supported by the National Science Foundation, Grant No. MC575-23738.*

ISBN 0-12-394680-8

INTRODUCTION

It has been estimated that the typical computer system spends over 25% of its time sorting - putting various files in order. In some environments, such as the insurance industry, the proportion is far higher. Sorting can greatly improve the efficiency of processing many different types of data, so this time is often very well spent. Literally hundreds of different sorting methods have been devised during the past thirty years, and the best of these have been subjected to extensive analysis. The performance of a sorting method is dependent on both the characteristics of the computer system on which it is run and the characteristics of the file to be sorted. It is difficult to determine which method is best for a particular application on a particular computer system, but many sorting methods are well enough understood that we can fairly accurately predict how they will perform.

Most computer manufacturers provide sorting programs as an integral part of the computer system, so that most programmers don't actually write new programs for particular sorting applications, but rather make use of the system provided, with little knowledge of the particular methods that it uses. This practice avoids needless duplication of effort, but it can lead to excessive costs. For example, consider the typical situation where files with similar characteristics are sorted on a regular basis. As the files grow (or their characteristics change in some other way), performance may degrade. Typically, performance is then enhanced by buying a bigger and faster computer. However, it may be that a sorting method better suited to the larger files might do much better. An understanding of how sorting methods work can lead to great savings in many situations.

This paper is intended to illustrate some of the basic mechanisms which are commonly used to sort by computer. We shall deal with methods appropriate for sorting files of all sizes from a few elements to a few million elements. While the methods described are not necessarily the best known, they are quite reasonable, and they are similar to methods in current use for practical situations. It is hoped that the elementary methods described will give the reader some appreciation for the kind of manipulations which must be performed in order to sort efficiently.

We shall deal throughout with a sample file chosen at random from the telephone book (see Fig. 1). Individual items in the file (also called *records*) consist of names, addresses, and telephone numbers. The goal is to sort the records to put them into alphabetic order by last name. The portion of the record used to determine the order is called the *key*.) In order to illustrate different methods, we shall make different assumptions about the capabilities of the computer system upon which the sort is to be performed. In all cases, we shall assume that the file is stored somewhere external to the computer system, say in a magnetic tape, and our goal is to produce another magnetic tape containing the sorted file.

INTERNAL SORTING: HEAPSORT

First, let us assume that the entire file will fit into the memory of our computer. Of course, our small sample file trivially satisfies this on most computer systems. The practical situations corresponding to this case are those for which we have a relatively large computer or relatively small files. Sorting methods for such files are called *internal* sorts, because we can read in the entire file, perform the sort internally and then output the sorted file.

It can be expensive to move large records around in a computer. A common way to avoid doing so is to perform a *pointer* sort, as follows. Suppose that while reading our records in, we

Jones Walter P 25 Bulck Point Av E Prov	433-4866
Wharf Tavern Water Wrn	245-5043
Morin Wilfred A dent 199 SMain Attl	222-3565
Briks Jos J 8 Buena Vista Ave SAttl	761-7233
Nanni Robt C Rawson Av Sfld	231-4153
Rhode Island National Guard Smithfield	231-0660
Alter Albert 67 Third St Prov	751-5024
Ted's Serv Sta 507 Washington Cov	821-9717
Imrie Gordon M Mrs 260 Wilson Av EProv	434-4913
Grain Everard B 114 Hilrd Av Wrwk	737-2607
Schur LM 317 Elm Natt	695-3346
Tetu Armand 61 Kent Cmb	728-4303
Ando Gaetano 231 Putnam Av Jstn	231-6591
Media Center Study 518 Lloyd Av Prov	421-3065
Spino Anthony J 33 Messina Prov	351-0170
Trofa Biagio 261 Lexingtn Av N. Prov	353-9154
Digby Chas L 26 Rialto Prov	861-7466
Wing Thos 76 Summit Ave Prov	331-2104
Pombo M L 440 Wickndn Prov	521-5564
Conte Chas N 360 Budlng Rd Crns	942-4381
Route 44 Gift Shop 1465 Atwood Ave Jstn	621-9796
Cason Norvel S 55 Ridgehill Rd Attl	222-9587
Beck Ernest R 900 Post Rd Wrwk	467-4809
Howdy Beef Burger 325 Warwck Av Wrwk	781-8989
Fair Jos D 4 Jean Dr Smfld	949-0664
Gince Geo Jr 66 N Garden Cmb	222-3546
McVan & Co 74 Falmth Attl	222-3546
Bebby Wm E 30 Warringtn Prov	781-9431
Leif Celia 66 Benefit Prov	351-3845
Mayo A H 165 Namcook Rd N Kgtwn	884-1205

FIGURE 1. Sample file to be sorted

assign an index to each one so that, in our example, Jones is assigned number 1, WHARF TAVERN number 2, Morin number 3, etc. Then, if we store these indices in a separate table, we can rearrange them rather than the records themselves. The records are always accessed through the indices. When the sort is complete, the index table for our example would be

7 13 28 23 4 22 20 17 25 26 10 24 9 1 29 30 27 14 3 5 19 6
21 11 15 8 12 16 2 18

indicating that the 7th record, Alter, should be the first output, then the 13th record, Ando, then the 28th record, Bebby, etc. This type of indexing is very easily programmed. From now on, we shall suppress the details involved by writing the keys in capital letters rather than the index numbers. After the records in our example have been read in, then, our index table is

JONES WHARF MORIN BRIKS NANNI RHODE ALTER TED'S IMRIE GRAIN
SCHUR TETU ANDO MEDIA SPINO TROFA DIGBY WING POMBO CONTE
ROUTE CASON BECK HOWDY FAIR GINCE MCVAN BEBBY LEIF MAYO

and our goal is to sort these into alphabetical order.

We shall assume that this index table is stored in an array, named A, and we shall use the notation A[i] to refer to the i-th element. For example A[6] refers to RHODE in our sample file. We shall denote the total number of records to be sorted by N, so that N = 30 in our example. The notation A[i:n] refers to all the elements between A[i] and A[n] (inclusive), so, A[1:N] refers to the entire table of N elements.

The method that we shall examine for internal sorting is a beautiful algorithm which involves structuring the array in a certain way, then using the structure to perform the sort. The data structure which is the basis for the algorithm is called a *heap*. An array A[1:N] is a heap if its elements satisfy the inequalities

$A[i] \geq A[2i]$

$A[i] \geq A[2i+1]$.

These inequalities, called the *heap property*, must hold for all i such that both sides specify array elements within A, specifically $1 \leq i \leq [N/2]$. Fig. 2 shows a heap containing our 30 names. To make the structure clear, it is convenient to draw heaps as trees, with lines from A[i] to A[2i] and A[ii+1], so that each element in the tree is bigger than both its successors.

We shall discover how to construct a heap from an arbitrary array after we have examined the problem of producing a sorted array from a heap.

It is not difficult to see how to begin. By the definition, A[1] must be the largest element in the array, so we know where it belongs in the sorted array, at A[N]. After exchanging A[1] and A[N] we no longer need to deal with A[N], but we also no longer have a heap. If we can easily transform A[1]...A[N-1] into a heap, then we can iterate the process to sort the array.

Figure 3a shows our sample heap after this step in the sorting process. Notice that the heap property is satisfied everywhere but at A[1]. To make the property hold at A[1], we simply exchange it with the larger of its two descendents (A[2], in the example), as shown in Figure 3b. Now the heap property holds everywhere except possibly at A[2]. Just as before, we make the property hold at A[2] by exchanging it with the larger of it's two descendents. (A[4] in the example) as in Figure 3c. After this exchange, we make property hold at A[4], and continue until A[1:N-1] is a heap.

In general, for an array of n elements, if all the elements in A[i+1:n] satisfy the heap property, then the following procedure will make all the elements in A[i:n] satisfy the heap property:

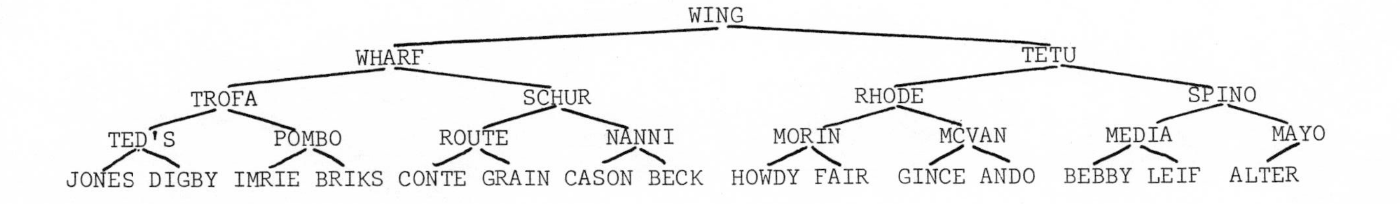

Figure 2. A heap

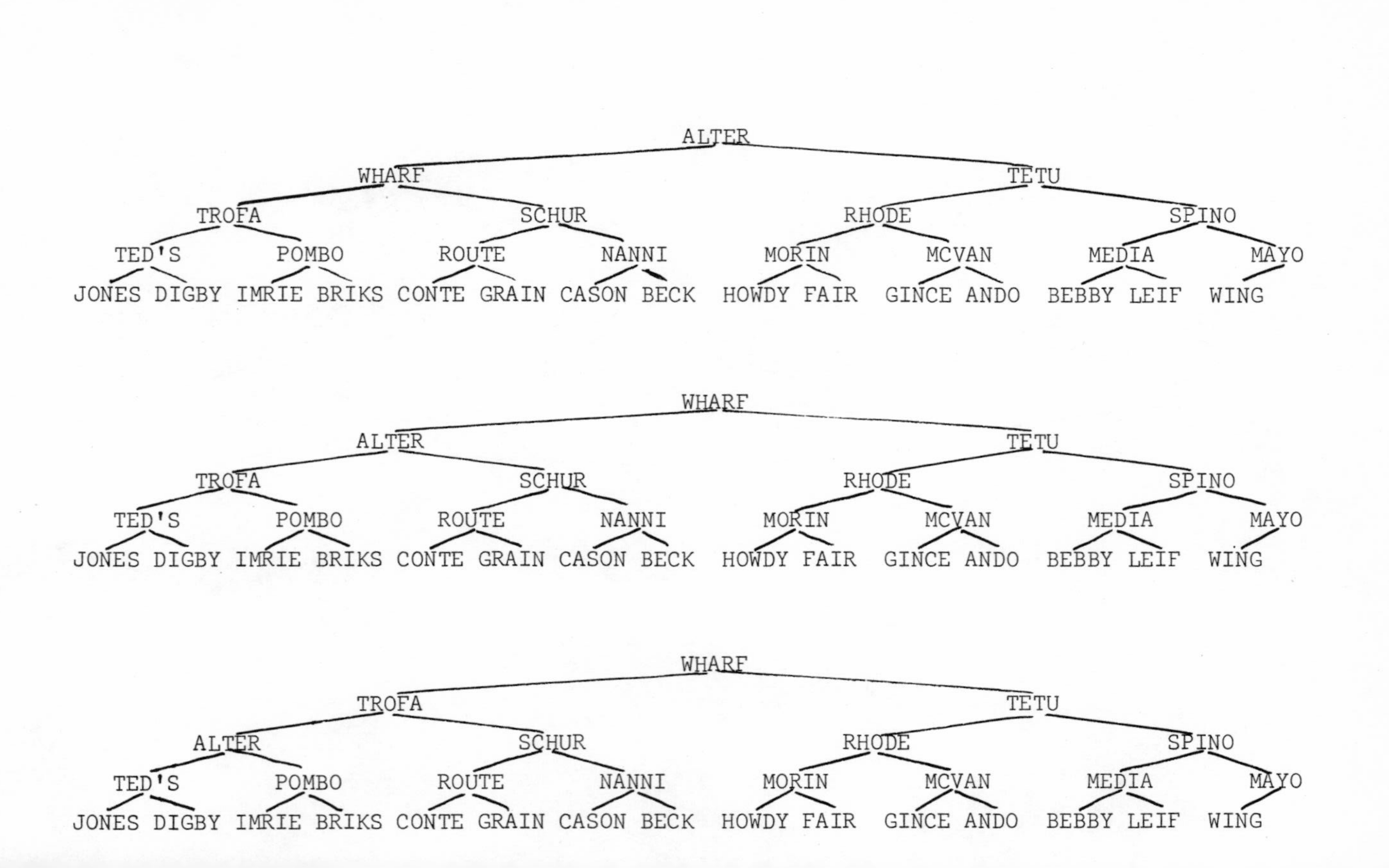

ALTER
WHARF TETU
TROFA SCHUR RHODE SPINO
TED'S POMBO ROUTE NANNI MORIN MCVAN MEDIA MAYO
JONES DIGBY IMRIE BRIKS CONTE GRAIN CASON BECK HOWDY FAIR GINCE ANDO BEBBY LEIF WING
WHARF
ALTER TETU
TROFA SCHUR RHODE SPINO
TED'S POMBO ROUTE NANNI MORIN MCVAN MEDIA MAYO
JONES DIGBY IMRIE BRIKS CONTE GRAIN CASON BECK HOWDY FAIR GINCE ANDO BEBBY LEIF WING
WHARF
TROFA TETU
ALTER SCHUR RHODE SPINO
TED'S POMBO ROUTE NANNI MORIN MCVAN MEDIA MAYO
JONES DIGBY IMRIE BRIKS CONTE GRAIN CASON BECK HOWDY FAIR GINCE ANDO BEBBY LEIF WING

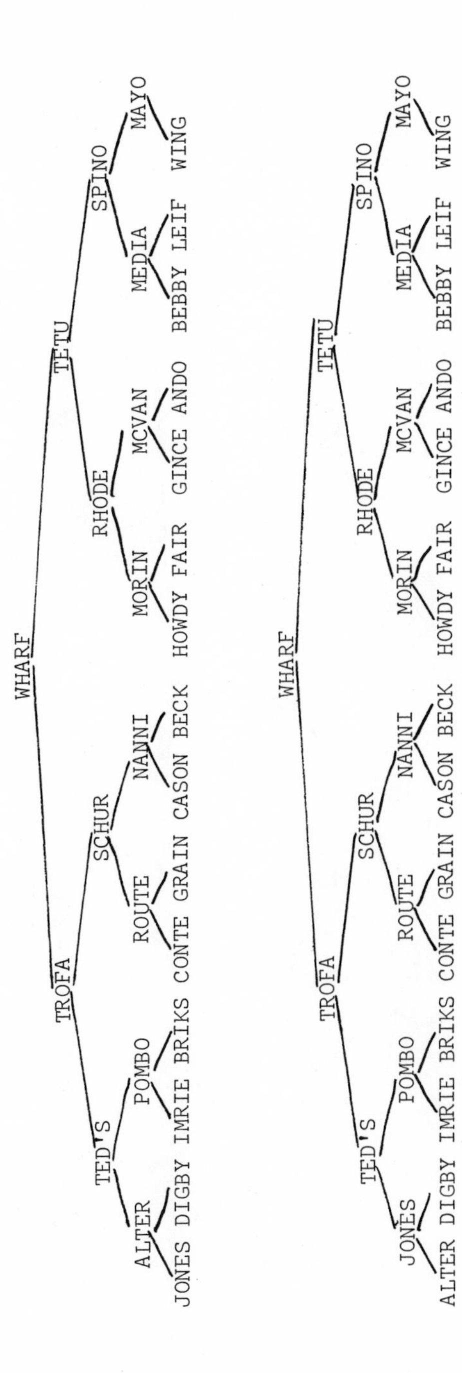

Figure 3. Reordering the heap

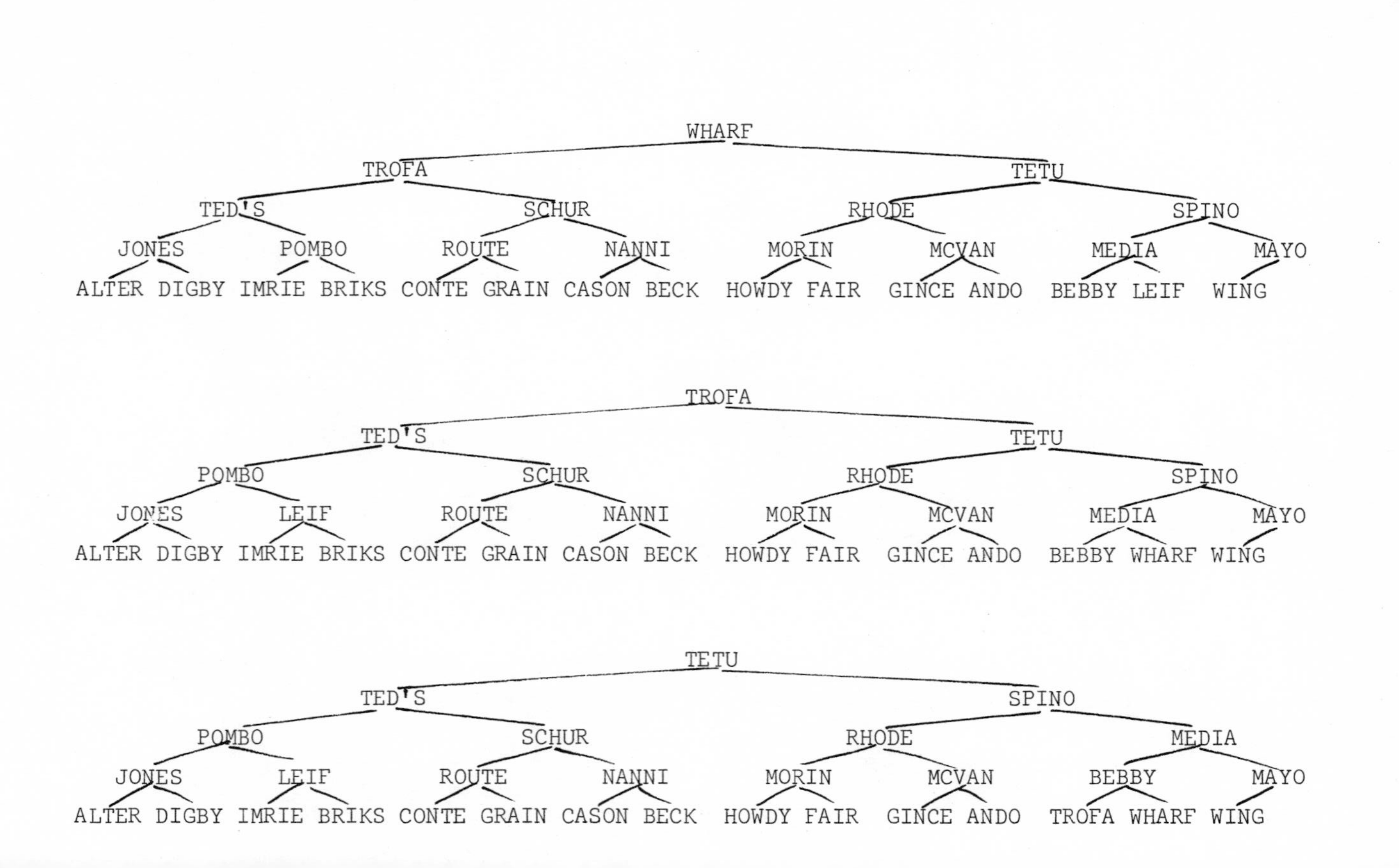

WHARF
TROFA TETU
TED'S SCHUR RHODE SPINO
JONES POMBO ROUTE NANNI MORIN MCVAN MEDIA MAYO
ALTER DIGBY IMRIE BRIKS CONTE GRAIN CASON BECK HOWDY FAIR GINCE ANDO BEBBY LEIF WING
TROFA
TED'S TETU
POMBO SCHUR RHODE SPINO
JONES LEIF ROUTE NANNI MORIN MCVAN MEDIA MAYO
ALTER DIGBY IMRIE BRIKS CONTE GRAIN CASON BECK HOWDY FAIR GINCE ANDO BEBBY WHARF WING
TETU
TED'S SPINO
POMBO SCHUR RHODE MEDIA
JONES LEIF ROUTE NANNI MORIN MCVAN BEBBY MAYO
ALTER DIGBY IMRIE BRIKS CONTE GRAIN CASON BECK HOWDY FAIR GINCE ANDO TROFA WHARF WING

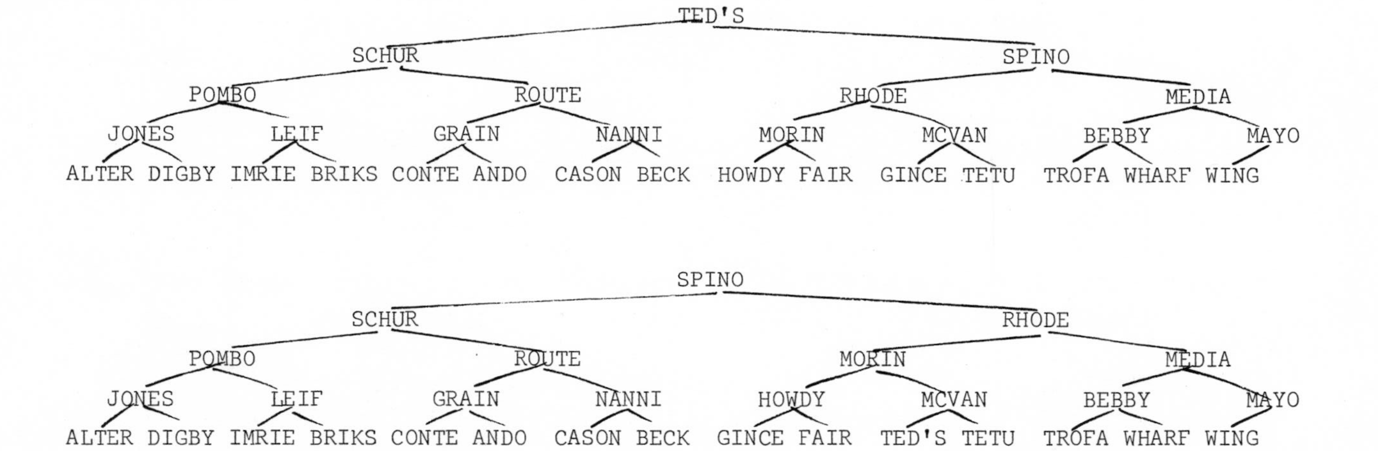

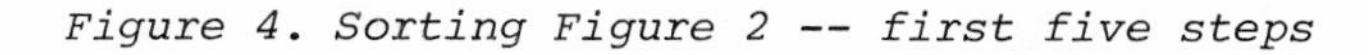
Figure 4. Sorting Figure 2 -- first five steps

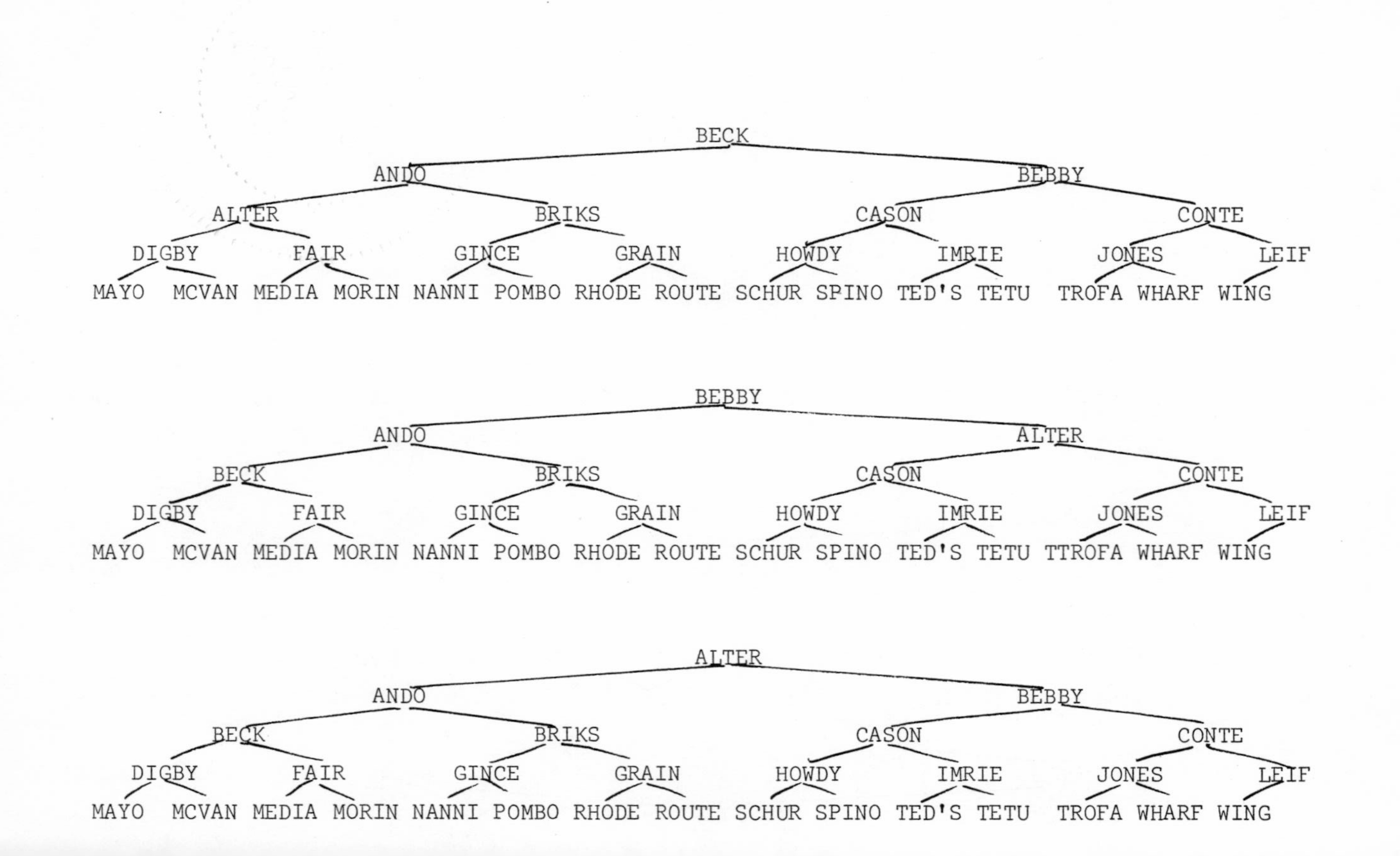
BECK
ANDO BEBBY
ALTER BRIKS CASON CONTE
DIGBY FAIR GINCE GRAIN HOWDY IMRIE JONES LEIF
MAYO MCVAN MEDIA MORIN NANNI POMBO RHODE ROUTE SCHUR SPINO TED'S TETU TROFA WHARF WING
BEBBY
ANDO ALTER
BECK BRIKS CASON CONTE
DIGBY FAIR GINCE GRAIN HOWDY IMRIE JONES LEIF
MAYO MCVAN MEDIA MORIN NANNI POMBO RHODE ROUTE SCHUR SPINO TED'S TETU TTROFA WHARF WING
ALTER
ANDO BEBBY
BECK BRIKS CASON CONTE
DIGBY FAIR GINCE GRAIN HOWDY IMRIE JONES LEIF
MAYO MCVAN MEDIA MORIN NANNI POMBO RHODE ROUTE SCHUR SPINO TED'S TETU TROFA WHARF WING

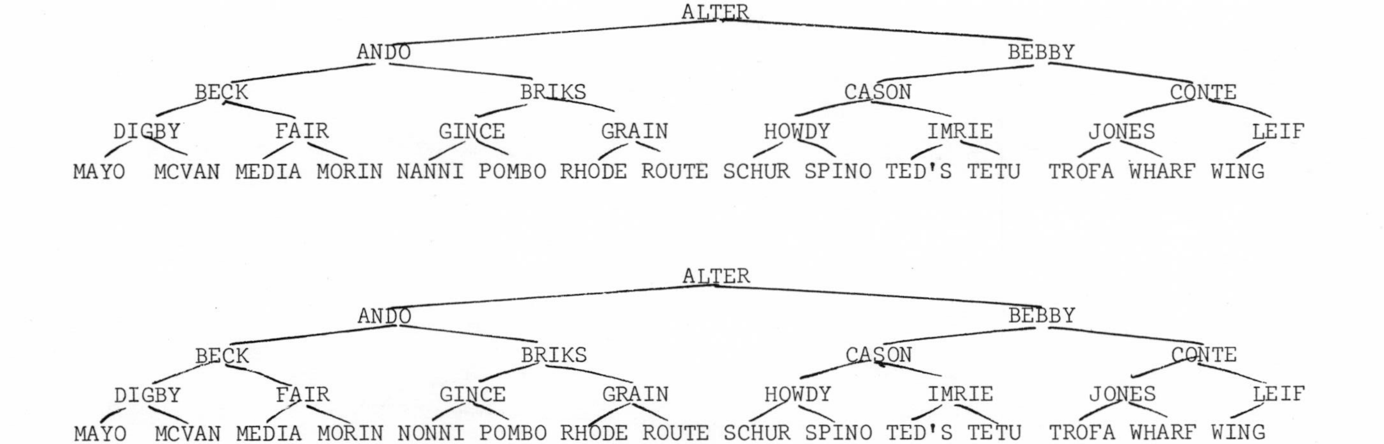

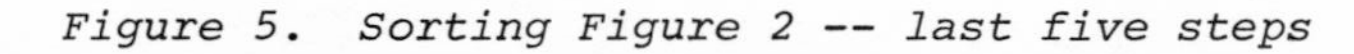

Figure 5. Sorting Figure 2 -- last five steps

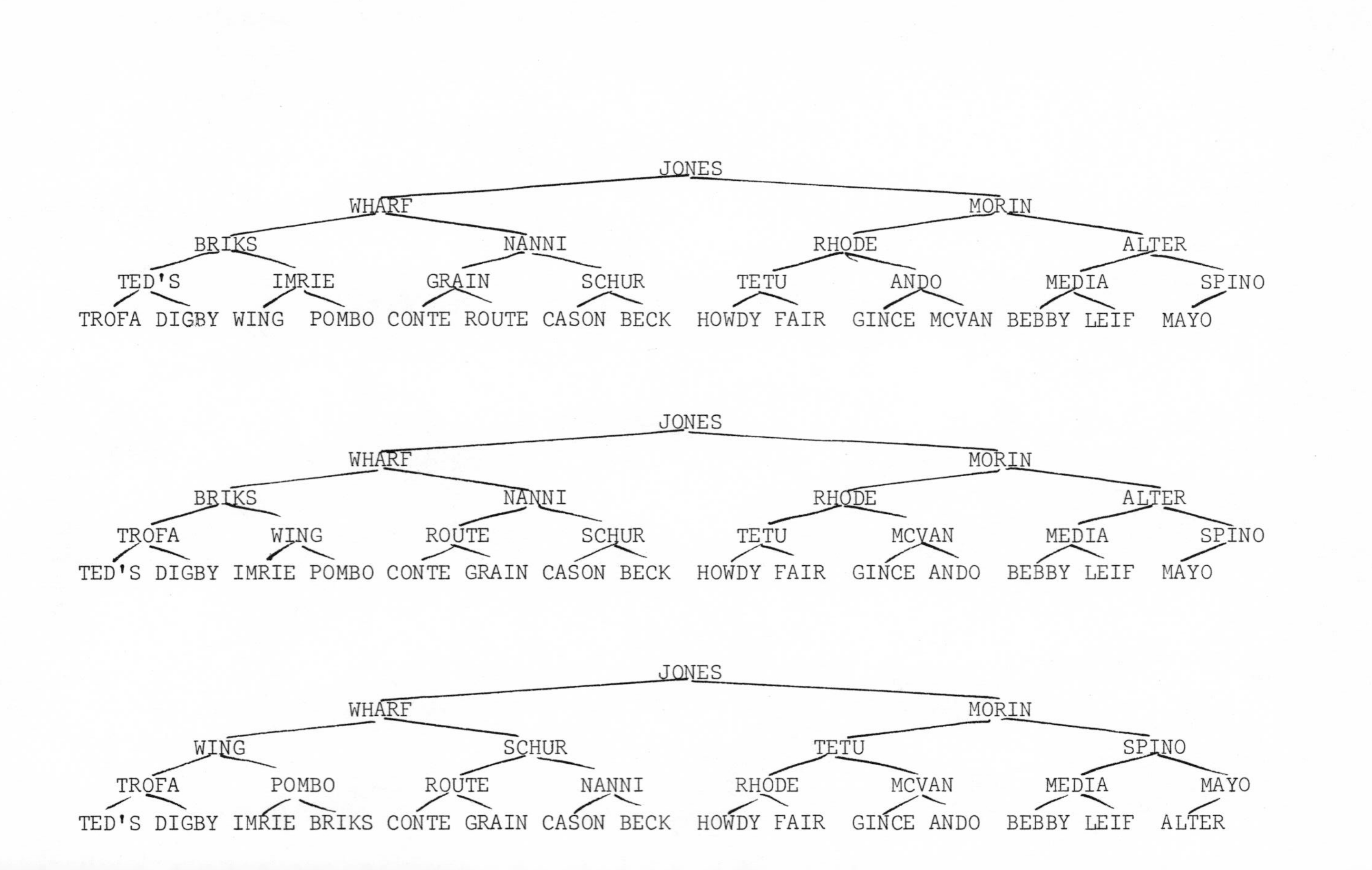
JONES
WHARF MORIN
BRIKS NANNI RHODE ALTER
TED'S IMRIE GRAIN SCHUR TETU ANDO MEDIA SPINO
TROFA DIGBY WING POMBO CONTE ROUTE CASON BECK HOWDY FAIR GINCE MCVAN BEBBY LEIF MAYO
JONES
WHARF MORIN
BRIKS NANNI RHODE ALTER
TROFA WING ROUTE SCHUR TETU MCVAN MEDIA SPINO
TED'S DIGBY IMRIE POMBO CONTE GRAIN CASON BECK HOWDY FAIR GINCE ANDO BEBBY LEIF MAYO
JONES
WHARF MORIN
WING SCHUR TETU SPINO
TROFA POMBO ROUTE NANNI RHODE MCVAN MEDIA MAYO
TED'S DIGBY IMRIE BRIKS CONTE GRAIN CASON BECK HOWDY FAIR GINCE ANDO BEBBY LEIF ALTER

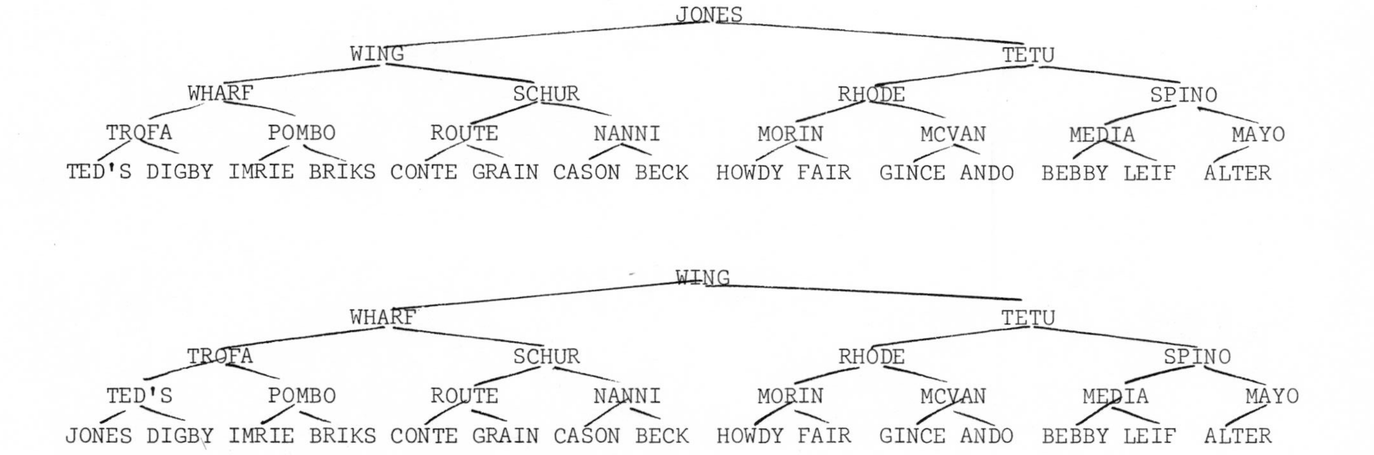

Figure 6. Building the heap

heaporder:

1. If i > [n/2] then stop. (A[i:n] satisfies the heap property).
2. If (2i+1) $\le$ n and A[2i+1] > A[2i], then set j to 2i+1, otherwise set j to 2i.
3. If A(i) < A(j) then exchange A[i] and A[j], set i to j, and return to Step 1.

With i set initially to 1 and n to N, this procedure works exactly as described above. Furthermore, from the discussion above, a heap can be transformed into a sorted array by using *heaporder* as follows:

for k = N,N-1,...,2, exchange A[1] and A[k], set i to 1 and n to k-1, and perform *heaporder.*

Figure 4 shows the first five steps in the sorting of our sample heap, and Fig. 5 shows the last five steps. Notice that the last drawing in Fig. 5, which is drawn as a tree, shows the array in order because of the way that the tree is stored.

If we can transform an arbitrary initial array into a heap, then we have a complete sorting method. Fortunately, we can use *heaporder* for this purpose also, by working backwards, from the bottom up:

for k = [N/2], [N/2] - 1, ..., 1, set i to k and n to N and perform *heaporder.*

Figure 6 shows how our sample heap was constructed by showing the situation after k = 7,3,1. Each time heaporder is performed the heap property is satisfied for A[k+1:N] so heaporder makes it satisfied for A[k:N]. After heaporder is performed with k = 1, then A[1:N] must be a heap.

The sorting method described above has been named *heapsort* by its inventor, J.W.J. Williams [12]. Williams' original method was improved to the method described above by R. W. Floyd [3]. The method has a quite distinguished history (see [6]), with its origins dating back to studies of lawn tennis tournaments by the Rev. C. L. Dodgson (Lewis Carrol).

Heapsort is widely used, for it has two important properties. First, it sorts inplace and does not require any extra "working storage". This is especially important in applications with large numbers of small records, when it is not necessary to do a pointer sort. Second, no matter how the input file is arranged, the running time of the method is guaranteed to be less than a small constant times $N \log_2 N$. This can be verified by counting the maximum number of exchanges taken by *heaporder*.

It is possible to show that the running time of *any* sorting method based on comparisons must be proportional to $N \log_2 N$, [8], so heapsort is, in this sense, nearly optimal. However, there are other methods which achieve this running time, and which can be expected to run faster than heapsort. The most prominent of these is a method called Quicksort [5,10], which will run twice as fast as heapsort on the average. Quicksort will take time proportional to N^2 in the worst case, but it is possible to make the worst case very unlikely, and the fast average performance makes Quicksort more useful than heapsort in practical applications. Other sorting methods may be more useful for some particular applications, but Quicksort and heapsort are the most prominent internal sorting methods.

EXTERNAL SORTING: REPLACEMENT SELECTION AND POLYPHASE MERGE

In many typical applications, the file to be sorted is much larger than will fit into the memory of the computer. Sorting methods designed to deal with such files are called *external* sorting methods.

Since the records will not all fit into the computer, it is necessary to pass through them several times, storing intermediate results on devices external to the computer. Another name that is often used for external methods is "sort/merge", for they almost all consist of two phases: a "sort" phase, where large runs of sorted records are produced, and a "merge" phase, where these are merged successively to produce the final sorted file.

External devices can be combined to produce a bewildering variety of computer systems, and the behavior of an external sorting method is very dependent on the particular system configuration being used. The technology is changing rapidly, and it is much harder to choose among the external methods than among the internal methods. External sorting involves implementing a sorting *system*, not merely one algorithm. Nevertheless, the problem has been studied extensively from an algorithmic point of view, and a number of interesting methods have surfaced.

To illustrate the issues involved, we shall treat the same example set of records as above, but we will assume that our computers memory is so small that it will hold only a few records (and pointers to them) at once. For external devices, we shall assume that we have three magnetic tape units, Tape 1, Tape 2, and Tape 3, with the unsorted input file on Tape 1. Magnetic tape devices are very slow relative to the rest of the computer, and good external sorting methods try to minimize input and output to the external devices. Tapes also have the property that they must be accessed serially: we can only read what is on the beginning of the tape. We shall briefly discuss methods appropriate for other types of external devices after examining the basic method.

First, let us consider the "sort" phase, where the goal is to pass our unsorted file through the computer, producing runs of sorted records. Suppose that our internal memory is only large enough to hold seven records. One way to proceed on our sample file of 30 records would be to read the records in seven at a time, sort them with heapsort, and write the sorted runs out.

This would produce five runs, four with seven records each and one (the last) with two records. It may seem that this naive approach is the best that we can do, but it is possible to produce runs that are much longer than can be held in the internal memory.

The idea is to take advantage of natural order in the input by processing the records one at a time, rather than in batches. Suppose that our memory is filled with the first seven input records:

JONES WHARF MORIN BRIKS MANNI RHODE ALTER.

The smallest of these is ALTER, so it clearly is the first record of the first run. After outputting the record associated with ALTER, we can immediately read in the next record to get

JONES WHARF MORIN BRIKS MANNI RHODE TED'S.

The new record, with key TED'S, is greater than ALTER, and can eventually be put in the same run. Continuing in this manner, we replace the smallest record in memory (BRIKS) by the next input record:

JONES WHARF MORIN IMRIE MANNI RHODE TED'S.

Now IMRIE is the smallest and it is replaced by GRAIN:

JONES WHARF MORIN GRAIN MANNI RHODE TED'S.

At this point, we must take further action, since GRAIN cannot belong to the first run (a larger key (IMRIE) has already been output). All we need do is mark it as belonging to the second run and proceed until the memory is full of keys belonging to the second run, at which point we begin forming successive runs in the same way. We are ready for a more formal description of this *replacement selection* method for the sort phase of external sorting:

Replacement selection:

1. Fill the memory, labelling all records with run number 1, and set r to 1.
2. Output the record with the smallest key among all records with run number r.
3. Read a new record into the place vacated. (If the input is exhausted, do nothing.) If the record just input has a smaller key than the record just output label it with run number r+1. Otherwise label it with run number r.
4. If no records are labelled with run number r, set r to r+1.
5. If no records remain, the sort phase is completed. Otherwise, go back to step 2.

The reader may find it instructive to verify that, when this method is applied to our sample file, it produces the three runs

1. ALTER BRIKS IMRIE JONES MORIN NANNI RHODE SCHOR SPINO TED'S TETU WHARF WING,
2. ANDO CASON CONTE DIGBY FAIR GINCE GRAIN HOWDY LEIF MAYO MCVAN MEDIA POMBO ROUTE,
3. BEBBY BECK.

We have not yet specified exactly how steps 2, 4 and 5, which involve testing all of the records in the memory, may be efficiently performed. Fortunately, as the careful reader will already have noticed, heaps are exactly the tool we need. If we maintain a heap with the inequalities reversed, then the *smallest* key will be on top, and Step 2 consists of outputting the top of the heap. Then Step 3 involves putting a new key on top of the heap and using *heaporder*. A convenient way to handle the effect of different runs is to append the run number for a record to the front of its key. (Thus, for example, 1. JONES will be less than 2. GRAIN but greater than 1. IMRIE.) The memory can be filled in a uniform way by first filling it with dummy records with run number 0, and it can be similarly emptied at the end by using dummy records with run number ∞ when the input is exhausted. These considerations lead to the following method for the sort phase.

Heap replacement selection:

1. Set A[1], ..., A[N] all to 0. DUMMY
2. Output the record associated with A[1].
3. If the input is exhausted, set A[1] to ∞. DUMMY. Otherwise, read a new record into the place vacated by the record just output and set the key field of A[1] to its key. If the key is smaller than the record just output, set the run number field of A[1] to r+1, otherwise set it to r.
4. Perform *heaporder* with i set to 1 and n to N (and with inequalities between keys reversed).
5. Set r to the run number field of A[1]. If r = ∞, the sort phase is completed. Otherwise, go back to Step 2.

The operation of this method on our sample keys is shown in Fig. 7 which depicts the heap just before each execution of step 2.

Analysis has shown that replacement selection can be expected to yield runs about twice as long as will fit in the internal memory [7]. Furthermore the method performs very well if there is order present in the input file. (This is a very common situation.) For example, if the input is already sorted, the method produces only one run. (Contrast this with the naive approach that we started with.) Another advantage of replacement selection is that, in a well designed system, input, processing, and output can be overlapped.

One problem that we have ignored in our discussion of the sort phase is the distribution of the runs on the output tape. One goal of the merge phase is to minimize "copying" of runs: runs should be read only as they are being merged with other runs, and written only as part of longer merged runs. Putting all the output runs on a single tape would involve immediate copying, so the runs need to be distributed among the available

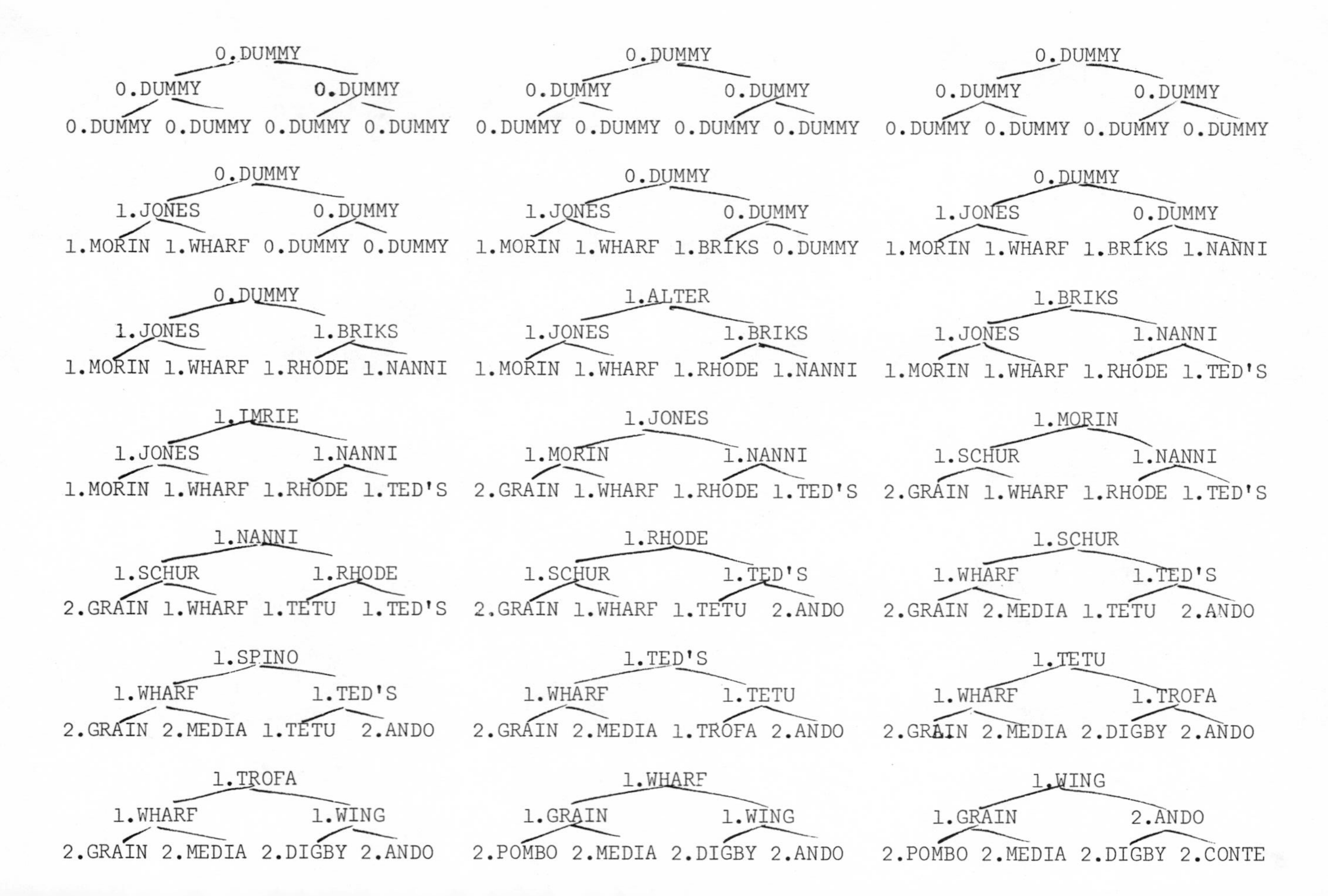
0.DUMMY
0.DUMMY 0.DUMMY
0.DUMMY 0.DUMMY 0.DUMMY 0.DUMMY
0.DUMMY
0.DUMMY 0.DUMMY
0.DUMMY 0.DUMMY 0.DUMMY 0.DUMMY
0.DUMMY
0.DUMMY 0.DUMMY
0.DUMMY 0.DUMMY 0.DUMMY 0.DUMMY
0.DUMMY
1.JONES 0.DUMMY
1.MORIN 1.WHARF 0.DUMMY 0.DUMMY
0.DUMMY
1.JONES 0.DUMMY
1.MORIN 1.WHARF 1.BRIKS 0.DUMMY
0.DUMMY
1.JONES 0.DUMMY
1.MORIN 1.WHARF 1.BRIKS 1.NANNI
0.DUMMY
1.JONES 1.BRIKS
1.MORIN 1.WHARF 1.RHODE 1.NANNI
1.ALTER
1.JONES 1.BRIKS
1.MORIN 1.WHARF 1.RHODE 1.NANNI
1.BRIKS
1.JONES 1.NANNI
1.MORIN 1.WHARF 1.RHODE 1.TED'S
1.IMRIE
1.JONES 1.NANNI
1.MORIN 1.WHARF 1.RHODE 1.TED'S
1.JONES
1.MORIN 1.NANNI
2.GRAIN 1.WHARF 1.RHODE 1.TED'S
1.MORIN
1.SCHUR 1.NANNI
2.GRAIN 1.WHARF 1.RHODE 1.TED'S
1.NANNI
1.SCHUR 1.RHODE
2.GRAIN 1.WHARF 1.TETU 1.TED'S
1.RHODE
1.SCHUR 1.TED'S
2.GRAIN 1.WHARF 1.TETU 2.ANDO
1.SCHUR
1.WHARF 1.TED'S
2.GRAIN 2.MEDIA 1.TETU 2.ANDO
1.SPINO
1.WHARF 1.TED'S
2.GRAIN 2.MEDIA 1.TETU 2.ANDO
1.TED'S
1.WHARF 1.TETU
2.GRAIN 2.MEDIA 1.TROFA 2.ANDO
1.TETU
1.WHARF 1.TROFA
2.GRAIN 2.MEDIA 2.DIGBY 2.ANDO
1.TROFA
1.WHARF 1.WING
2.GRAIN 2.MEDIA 2.DIGBY 2.ANDO
1.WHARF
1.GRAIN 1.WING
2.POMBO 2.MEDIA 2.DIGBY 2.ANDO
1.WING
1.GRAIN 2.ANDO
2.POMBO 2.MEDIA 2.DIGBY 2.CONTE

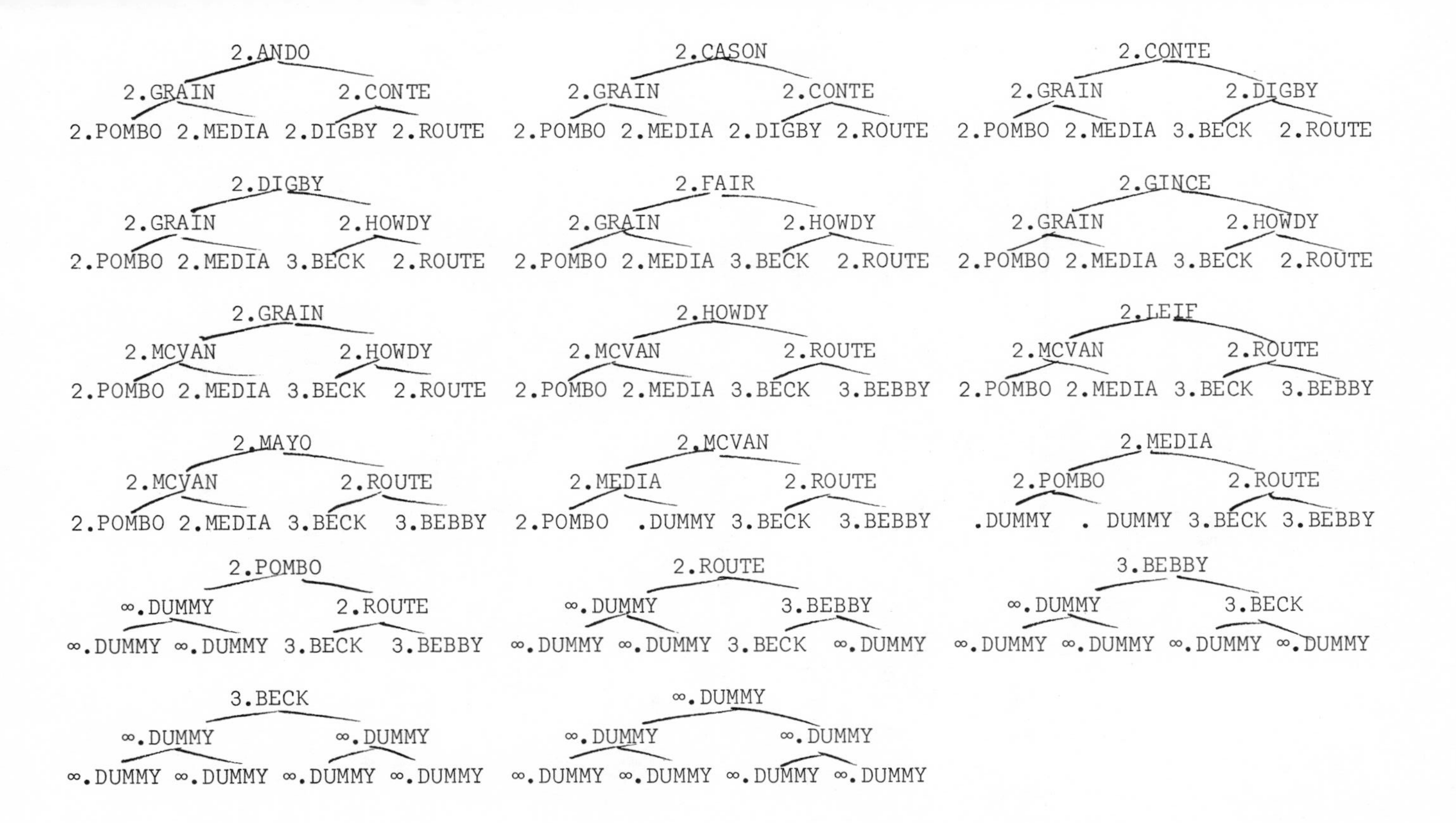

Figure 7. Heap replacement selection, N=7

tapes to set things up for the merge phase. If there are only a few runs, it is clear what to do. For example, on our three-tape system, if there are only two runs, we can put one on Tape 2 and one on Tape 3, so the merge phase consists of simply merging these to produce the sorted file back on Tape 1. If there are three runs (as in the example above), we can put two on Tape 2 and one on Tape 3. Then there are two merging stages: first merge the run on Tape 3 with the first run on Tape 2 to put one (longer) run on Tape 1; then merge this with the remaining run on Tape 2 to put the sorted file back on Tape 3. For large numbers of runs, things become much more complicated, and a wide variety of distribution and merging strategies have been proposed. We shall continue our discussion of the distribution problem after examining one of the fundamental merging methods, the *polyphase merge*.

To get more runs to work with in our example, let us suppose that our internal memory is so small (or our records so large) as to hold only two records at most. As we shall see, we can sort the file quite efficiently even under this severe constraint. If the replacement selection method described above is applied to our sample file for N = 2, we get the eight runs

1. JONES MORIN WHARF
2. BRIKS NANNI RHODE TED'S
3. ALTER GRAIN IMRIE SCHUR TETU
4. ANDO MEDIA SPINO TROFA WING
5. DIGBY POMBO ROUTE
6. CASON CONTE HOWDY
7. BECK FAIR GINCE MCVAN
8. BEBBY LEIF MAYO.

It turns out that the best way to distribute eight runs on our three tape units is to put five on one unit and three on another. (We shall see why soon.)

Once the runs have been produced and distributed, the polyphase method is very easily described:

Polyphase merge:

1. Runs must be distributed among the tapes according to the pattern described below. One tape must be initially empty. Call this the *output* tape and the others *input* tapes.
2. Merge the first runs on the input tapes to the output tapes. Repeat this step until an input tape becomes empty.
3. If the whole file is on the output tape the sort is complete. Otherwise rewind the output tape (which now becomes an input tape), designate the empty input tape as the new output tape, rewind it, and go back to Step 1.

(To "rewind" a tape is to position it at the beginning, which can be done at high speed.) Figure 8 shows the contents of the tapes for our sample file just before each execution of Step 2. (A full cycle of "merge until empty" step is called a *phase* - hence the name "polyphase".) As mentioned above, we start with five runs on Tape 2 and three runs on Tape 3. After the first phase, we have two runs left on Tape 2 and three (longer) runs on Tape 1. Then the second phase leaves one run on Tape 1 and two (even longer) runs on Tape 3, and the merge is completed with the three-run case described above.

The trick to deciding how the initial runs should be distributed for a general sort is to work backwards to build a table describing the number of runs on each tape at each phase. We know that at the end we want only one run (the sorted file) on only one of the tapes; say Tape 1, with Tape 2 and 3 empty. We can describe this distribution with the triplet:

1 0 0.

This run had to be produced by a merge involving both other

Tape1	Tape2	Tape3
	JONES	BRIKS
	MORIN	NANNI
	WHARF	RHODE
	ALTER	TED'S
	GRAIN	ANDO
	IMRIE	MEDIA
	SCHUR	SPINO
	TETU	TROFA
	DIGBY	WING
	POMBO	CASON
	ROUTE	CONTE
	BECK	HOWDY
	FAIR	
	GINCE	
	MCVAN	
	BEBBY	
	LEIF	
	MAYO	

Tape1	Tape2	Tape3
BRIKS	BECK	
JONES	FAIR	
MORIN	GINCE	
NANNI	MCVAN	
RHODE	BEBBY	
TED'S	LEIF	
WHARF	MAYO	
ALTER		
ANDO		
GRAIN		
IMRIE		
MEDIA		
SCHUR		
SPINO		
TETU		
TROFA		
WING		
CASON		
CONTE		
DIGBY		
HOWDY		
POMBO		
ROUTE		

Tape1	Tape2	Tape3
CASON		BECK
CONTE		BRIKS
DIGBY		FAIR
HOWDY		GINCE
POMBO		JONES
ROUTE		MCVAN
		MORIN
		NANNI
		RHODE
		TED'S
		WHARF
		ALTER
		ANDO
		BEBBY
		GRAIN
		IMRIE
		LEIF
		MAYO
		MEDIA
		SCHUR
		SPINO
		TETU
		TROFA
		WING

Tape1	Tape2	Tape3
	BECK	ALTER
	BRIKS	ANDO
	CASON	BEBBY
	CONTE	GRAIN
	DIGBY	IMRIE
	FAIR	LEIF
	GINCE	MAYO
	HOWDY	MEDIA
	JONES	SCHUR
	MCVAN	SPINO
	MORIN	TETU
	NANNI	TROFA
	POMBO	WING
	RHODE	
	ROUTE	
	TED'S	
	WHARF	

Tape1	Tape2	Tape3
ALTER		
ANDO		
BEBBY		
BECK		
BRIKS		
CASON		
CONTE		
DIGBY		
FAIR		
GINCE		
GRAIN		
HOWDY		
IMRIE		
JONES		
LEIF		
MAYO		
MCVAN		
MEDIA		
MORIN		
NANNI		
POMBO		
RHODE		
SCHUR		
SPINO		
TED'S		
TETU		
TROFA		
WHARF		
WING		

Figure 8 Polyphase merge

tapes, so at the beginning of the previous phase the distribution must have been

0 1 1.

Now, suppose that Tape 2 was the output tape for the phase previous to this. Its one run must have been produced by a merge from Tapes 1 and 3, and the previous distribution must have been

1 0 2.

Continuing, we pick the tape with the most runs as the "output" tape, and the previous run distribution must have been

3 2 0.

The next is

0 5 3

which is the distribution that we used in our example, and we can continue in this manner to get Fig. 9, the table of "perfect" polyphase run distributions. If we have 1,2,3,5,8,13,21,34,55, 89,144,233 runs to merge, this table shows exactly how to distribute the runs and how the merge proceeds.

Some readers may recognize the numbers which are omnipresent in Fig. 9. They are the well-known Fibonacci numbers, which are defined by the formula

$$F_{n+2} = F_{n+1} + F_n \quad n \geq 0, \text{ with } F_0 = 0 \text{ and } F_1 = 1.$$

These numbers have been studied very extensively, as they arise in a variety of natural situations (not to mention a number of computer algorithms). A simple inductive proof from Fig. 9 tells us that, if we have F_n runs to be merged, then we should distribute F_{n-1} onto one tape and F_{n-2} onto another and use polyphase. The sort will then require n-2 phases.

In a real sorting application, we may not have exactly F_n runs to be merged. What's worse, we don't know in advance how many runs there are, if we are using replacement selection. The

Tape1	Tape2	Tape3	Total Runs
1	0	0	1
0	1	1	2
1	0	2	3
3	2	0	5
0	5	3	8
5	0	8	13
13	8	0	21
0	21	13	34
21	0	34	55
55	34	0	89
0	89	55	144
89	0	144	233

Figure 9. Polyphase numbers

usual practice is to distribute the runs equitably during the sort phase, then use "dummy" runs to fill in up to the perfect distribution. In our three-tape example, we can keep count during the sort phase of the total number of runs output so far, and switch tapes each time the count reaches a Fibonacci number.

The polyphase method just described for three tapes was apparently first discovered by B. K. Betz [1]. If more tapes are available, it is possible to develop a pattern based on generalized Fibonacci numbers by working backwards as above [4]. The method has been subjected to extensive analysis, and some very interesting theoretical results have been developed. For example, D. A. Zave has recently shown how dummy runs may be distributed in an optimal manner [13].

From a practical standpoint, however, polyphase has a wide variety of competitors. A similar pattern called cascade merge will do better if six or more tapes are available [1]. Other methods take into account the rewind time, or other special properties of particular tape units. It is also possible to gain efficiency by interleaving the sorting and merging phases in an "oscillating" sort [11]. If other types of external devices are available, such as disks, then so-called "balanced" merging method methods may be appropriate (see [6]). Each technological advance leads to a new family of methods, and comparison of the various external methods is not nearly as clear-cut as comparison of internal methods.

CONCLUSION

Methods for sorting by computer have only been under study for about thirty years, but a remarkable variety of different algorithms have been proposed. Different methods have been studied extensively through theoretical analysis, empirical experimentation, and performance measurement in practical applications. The problem of choosing a method appropriate for a particular application on a particular computer is much less difficult than it once used to be. Especially in the case of internal sorting, we are able to draw some rather definite conclusions about the relative performance of the various methods.

The algorithms that we have examined are representative of many of the methods which have been developed. Through these few algorithms, readers may have some appreciation of what it takes to make a computer sort a file. Readers interested in learning other methods or in studying sorting more deeply should consult Knuth's book on the subject [6], and the related bibliography [9] covers most of the important literature on the subject.

REFERENCES

[1] Betz, B. K. and Carter, W. C. New merge sorting techniques. Paper 14, preprints of summaries of papers, 14th Natl. Meeting of the ACM, Cambridge, Mass., 1959, 4 pp.

[2] Dinsmore, R. J. Longer strings from sorting. *Comm. ACM, 8, 1* (Jan. 1965), p. 48.

[3] Floyd, R. W. Treesort 3 (Algorithm 245) *Comm. ACM 7, 12* (Dec. 1964), p. 701. (See also certifications by D. S. Abrams in *Comm. ACM 8, 7* (July 1965), p. 445 and by R. L. London in *Comm. ACM 13, 6* (June 1970), pp. 371-373.)

[4] Gilstad, R. L. Polyphase merge sorting-an advanced technique. Proc. Eastern Joint Computer Centr. *18*, 1960, Spartan Books, New York, pp. 143-148.

[5] Hoare, C.A.R. Quicksort. *Computer J. 5* (April 1962), pp. 10-15.

[6] Knuth, D. E. "Sorting and Searching," The Art of Computer Programming *3*, Addison-Wesley, Reading, Mass., 1973.

[7] Moore, E. F. Sorting method and apparatus. U. S. Patent #2,983,904, May 9, 1961.

[8] Morris, R. Some theorems on sorting. *SIAM J. Appl. Math. 17* (1909), pp. 1-6.

[9] Rivest, R. L. and Knuth, D. E. Computer Sorting (Bibliography 26). *Comp. Reviews 13, 6* (June 1972), pp. 283-289.

[10] Sedgewick, R. Quicksort. Ph.D. Thesis, Stanford University, Stanford, Calif., 1975 (Stanford CS Report).

[11] Sobel, S. Oscillating sort - a new sort merging technique. *J. ACM 9* (1962), pp. 372-374.

[12] Williams, J.W.J. Heapsort (Algorithm 232) *Comm. ACM 7, 6* (June 1964), pp. 347-348.

[13] Zave, D. Optimal polyphase sorting. *SIAM J. Comp.* (to appear).

SYMBOLIC INFORMATION PROCESSING

Kenneth M. Levine

Mutual Life Insurance Company of New York
New York, New York

Peter Masters

Insurance Services Office
160 Water Street
New York, New York

I. ANALYSIS OF ALGORITHMS

An elaboration of the distinction between marco-evaluation time, compile time, and execute time, leading to a notion of a hierarchy of times and places. Significant improvement in program performance.

Sometimes the obvious solution is not the best one.

Symbolic Information Processing is born of looking at problems in an algebraic, rather than arithmetic, manner. When Gauss was ten years old, his arithmetic teacher asked his pupils to add up all the whole numbers from one to one hundred. Gauss wrote a single number, five thousand and fifty, on his slate, while the other students struggled with the sum. He had chosen to solve the

ISBN 0-12-394680-8

trivial algebraic problem. One cannot imagine that he was not aware, at the moment he wrote the answer, that he knew how to find the sum of any arithmetic series.

This process of abstraction, from object to symbol, underlies every theoretical model of computation. The Turing machine paradigm is not the only one. Other equally valid, because equally powerful, formalisms exist; some exhibit more algebraic behavior than others. Although literally isomorphic to Turing's machine, Church's λ-calculus is aesthetically of a different nature.

Aesthetics make a difference, because the human mind can comprehend so little of the (information-theoretic) content of a large computer program.

We present the analysis of an algorithm that differentiates polynomials, in the hope that we can convey to you something of what Gauss must have felt that day in Braunschweig.

The problem is as follows: We are given a polynomial or rational function in x, specifying a function we may call f(x). We are to produce a polynomial formula in x, specifying the function f'(x).

For instance, if we are given the function

x ↑ 2

we are expected to return its derivative

2x.

This problem may not be of the sort you are used to dealing with. It involves the manipulation of symbols, not their evaluation. We all learned, in freshman calculus, an algorithm for differentiating rational functions of any degree of complexity. For example, let

$$y = \left(\frac{4x+3}{x}\right)^2 \cdot \left(\frac{5x^2-9}{6x}\right)$$

How do we evaluate y'? First, we note that we have a product of two algebraic polynomials in x. If we can differentiate both of them, then we can use

if $y = f(x).g(x)$
then $y' = f(x).g'(x) + g(x).f'(x)$

to compute y'. Then, we have two subproblems to solve. The first is to find $\frac{d}{dx}\left(\frac{4x+3}{x}\right)^2$. By a similar argument, we reduce this to the problems,

$$\frac{d}{da}(a^2),\ \frac{d}{dx}\left(\frac{4x+3}{x}\right),\ \text{and}\ \frac{d}{dx}f(g(x)) = \frac{df}{dg}\cdot\frac{dg}{dx}.$$

We use the quotient formula,

$$\text{if } y = \frac{f}{g},\ \text{then } y' = \frac{g'.f-f'.g}{g.g}$$

to find the derivative of $\left(\frac{4x+3}{x}\right)$ in terms of $\frac{d}{dx}(4x+3)$ and $\frac{d}{dx}c$.

Formulas for the sum and product, together with the formulas

$$\frac{dc}{dx} = 0 \quad \text{and} \quad \frac{dx}{dx} = 1$$

are sufficient to solve the problem.

This way of solving the problem made use of the way we think of the formula, as a complex of more simple formulas. A description of the formula, that makes explicit (to the program) how the formula is constructed, will make much easier the definition of a differentiation program.

Our first impression of our sample formula is that it is a product of two terms. So, at the very beginning, we want to look at our formulas as

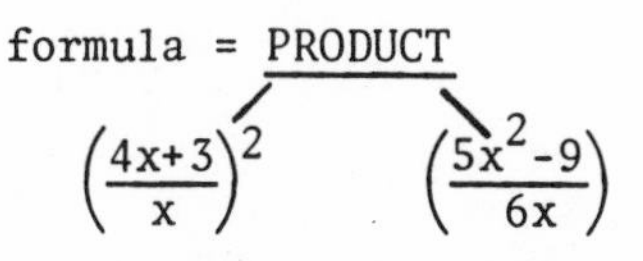

Now, examine the left-hand term. It is the square of a "simpler" formula

```
= SQUARE
     |
  (4x+3)
  ( ---- )
  (  x   )
```

and that formula is

```
   = QUOTIENT
     /      \
(4x+3)      (x)
```

Each of these forks is an explanation, or a translation, of a previous term. We can put them all together, to get a representation of our data structure:

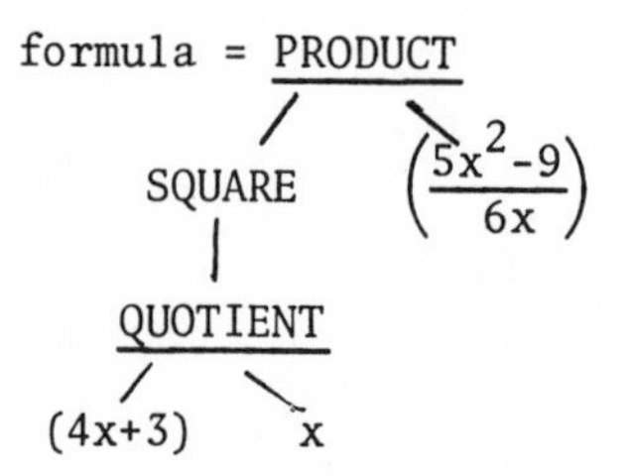

Clearly, we can extend the process until our only formulas left unevaluated are x and constants. We can replace (4x+3) with

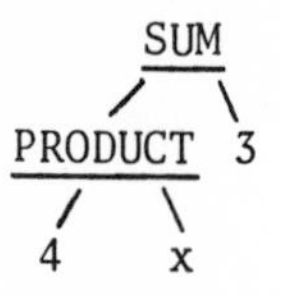

Suppose we can represent all this symbolic information in a data variable - a variable, we note, whose value is a tree (call this variable E). Then, it will be a simple matter to define an algorithm to differentiate e:

Evaluate the first formula that applies:

1. If E is a constant, its derivative is zero.
2. If E is the monomial "x", its derivaite is 1.
3. If E is the sum of two terms, its derivative is the sum of their derivatives.
4. If E is the difference of two terms, its derivative is the difference of their derivatives.

5. If E is the product of two terms, A and B, then the derivative of E is A times the derivative of B, plus B times the derivative of A.
6. If E is the quotient of two terms, A and B, then the derivative of E is B times the derivative of A, minus A times the derivative of B, all divided by the square of B.
7. If E is a term A raised to a constant power n, then the derivative of E is equal to n times A, raised to the (n-1) power, times the derivative of A.

Since the expressions differentiated in each call to the algorithm are simpler than the preceeding expressions, we know the procedure will halt. It is indeed the algorithm we need; all that is left is to find a good way of storing data as trees.

The language we will use to code our algorithm is LISP, a language developed to solve problems of a symbolic nature.[1] Let us begin, by writing out the structure we will use to represent our sample formula. The value of E is:

```
PRODUCT
  (POWER
     (QUOTIENT
        (SUM
           (PRODUCT 4  X
           3)
        X)
     2)
     (QUOTIENT
        (DIFFERENCE
           (PRODUCT
              5
              POWER X  2))
          9)
       (PRODUCT 6  X) ))
```

[1]*See (Weissman 1967) and (McCarthy 1965) for LISP. Our example is drawn from Weissman, Chapter 20. All references in parentheses are to be directed to the Bibliography.*

This is a data value in our program. None of the words has its normal meaning, as a LISP function name. We can think of the words as being symbols, or character strings, in our data base, which our program can recognize. It is clear that we can print this data element in a way that makes it look like a tree; a moment's reflection will show we can tell how to print it this way, by watching parentheses.

LISP has only a few primitive data types. One is the ATOM, the smallest "point structures" from which the others are built. ATOMs include numbers, such as 4, and literal atoms (identifiers or variables) such as X. The other common LISP data type is the LIST.

A LIST may be thought of as a sequence of LISP data values, "enclosed in parentheses". Thus, (A 2 B) is a LIST, composed of the three elements, A, 2, and B. What gives this formalism power is that any LISP value, including LISTs, can be an element of a list. Thus, we can have lists like

```
     (A (B D) Q)
or   (A Q (B D)).
```

These two lists, by the way, are different. The language is equipped to extract the first element of a LIST, or the LIST without its first element (this reflects the internal representation of LISTs in LISP. For this reason, a LIST is not to be considered as a variable length vector). These two functions are enough to find us any specified element of a LIST. The second element of a LIST, is the first element of the rest of the list, the

```
     first element(rest(LIST))
```

or, in LISP terminology,

```
     (CAR (CDR LIST))
or,  (CADR  LIST).
```

CAR and CDR are the traditional LISP names for "take the first element" and "take the rest of the list." Compound symbols, such as CADR or CADADDDR, may be interpreted by taking every A as "take the CAR of" and every D as "Take the CDR of." Thus, the CADR of ((X Y)(A B)) is (A B), and the CAADR is A.

Going back to our main example, we find the list looks like this:

```
E = (PRODUCT  list1  list2)
```

So we would like to code step 5 of our algorithm:

```
     If the CAR of E is PRODUCT, then compute the derivative
of E as follows:
     We need the first term, f(x), which is the CADR of E.
Similarly, g(x) is the CADDR of E.
     We want to construct from these, the formula
f.g' + g.f', that is
                   (SUM
                      (PRODUCT
                         (CADR  E)
                         (DERIV  (CADDR  E)  ))
                      (PRODUCT
                         (CADDR  E)
                         (DERIV CADR E)  )))
```

Of course, we've only discussed half of the language; the data types. We have still not defined what a program statement looks like. Nor will we; but we will tell you a little bit about what's going on.

The first problem we get out of the way, is variables and constants. If "X" can be a constant in our program, then what can we use for variables? The answer, in LISP, is to have it both ways. Every identifier is assumed to be a variable, unless it is QUOTEd. 'X is QUOTEd, X is not.

Every statement in LISP is a list. We interpret or evaluate a LISP statement as follows: the first element (the CAR) is the name of a function. The rest of the elements are this function's arguments. Thus, (SQRT 4) evaluates to 2. (PRINT E) causes the value of the variable E to be printed. (PRINT 'E) causes the letter E to be printed.

LIST is a function, which forms a list of all its arguments. So (LIST 'A 'B '(C D)) evaluates to (A B (C D)).

So, we now know better how to specify the part of step 5 which builds the list to represent f.g' + g.f'. We write:

```
(LIST 'SUM
   (LIST 'PRODUCT
      (CADR E)
      (DERIV (CADDR E) ))
   (LIST 'PRODUCT
      (CADDR E)
      (DERIV (CADR E) )))
```

Obviously, we have not covered all the aspects of LISP necessary to program in it. Nonetheless, we will present the complete program to differentiate algebraic polynomials (represented as a list structure). A few hints:

LAMBDA is LISP's SUBROUTINE statement, (E) is its parameter list.
(ATOM E) is true (T) if E is an ATOM.
(EQ A B) is true (T) if A equals B.
(EQUALS A B) is true (T) only if A and B are the same location.
(COND (A B)(C D) means
    If A is true, then evaluate or perform B. Else, if C is true, then perform D.
NIL and () are synonyms.
T is always true.
ADD, SUBTRACT, MULTIPLY, and DIVIDE are themselves.

The function we are writing is named DERIV. Its value is:

```
LAMBDA  (E)                             ;if E is an atom, then
  (COND  ((ATOM  E)                     ;if E is X return 1, and
     (COND  ((EQ  E  X)  1)(T  0) ))    ; if E is not X, return 0.

     ((OR  (EQ  (CAR  E)  'SUM)         ;if E is a sum
           (EQ  (CAR  E)  'DIFFERENCE)) ;or a difference
      (LIST  (CAR  E)                   ;return the sum or the
         (DERIV  (CADR  E))             ;difference of the derivative
         (DERIV  (CADDR  E)) ))         ;of the terms

      ((EQ  (CAR  E)  'PRODUCT)         ; this is just the code we
       (LIST  'PLUS                     ;previously saw for
          (LIST  'PRODUCT  (CADR  E)    ;f.g' + g.f'
             (DERIV  (CADDR  E) ))
          (LIST  'PRODUCT  (CADDR  E)
             (DERIV  (CADR  E) )))

       ((EQ  (CAR  E)  'QUOTIENT        ;the rule for quotients
        (LIST  'QUOTIENT
           (LIST  'DIFFERENCE
              (LIST  'PRODUCT
                 (CADDR  E)

             (DERIV  (CADR  E) ))       ;the rule for
           (LIST  'PRODUCT              ;exponentiation.
             (CADR  E)
             (DERIV  (CADDR  E) )))
         (LIST  'PRODUCT  (CADDR  E)(CADDR  E)) ))
```

```
        ((EQ  (CAR  E)  'POWER)
         (LIST  'PRODUCT
            (LIST  'PRODUCT  (CADDR  E)
               (COND  ((EQUAL  (CADDR  E)
                         2)
                      (CADR  E))
                     (T  (LIST  'POWER
                           (CADR  E)
                           (SUB1  (CADDR  E))
                                )) ))
            (DERIV  (CADR  E)) ))

        (T NIL)) ))                              ;We must have a pair
                                                 ;which is evaluated,
                                                 ;to make COND happy.
```

The code is twenty lines long, in Weissman's original presentation. It is almost characteristic of a LISP program, that our final statement so closely resembles the algorithm we defined earlier. Most languages, that could do this program at all, would exhibit this structure in their IFs and DOs; LISP is unique in the extent to which this structure is all there is to the program. If one defined the function SUM by

```
    (LAMBDA  (A  B)
       (LIST  'SUM
          (DERIV  A)
          (DERIV  B)  ))
```

and similarly for DIFFERENCE, PRODUCT, QUOTIENT, and POWER, then, if DERIV were

```
    (LAMBDA  (E)
       (COND  ((ATOM  E)
          (COND  ((EQ  X  E)  1)(T  0)  ))
          (T  (EVAL  E))  ))
```

we would get our result, by treating our data as a program, and by evaluating it. DERIV gives us an evaluation environment, or context, in which "the value of X is 1; the value of any other atom is 0".

It may not be obvious that the tree can claim anything close to universal applicability to our problem-solving. We remind you, however, of the very first step we took: How do we evaluate the derivative of a product? The answer, we realized, is that we break the problem up into smaller sub-problems:

> since a product is the product of two subexpressions we, first, find each of their derivatives, then apply our product formula once.

This is probably about as much strategy as we would ever apply in solving a problem, *at any one time*. When we start thinking about one of the sub-expressions, we forget about what we've been doing, and concentrate on the sub-problem at hand. LISP lets us do this, painlessly.

Since each task, from the top down, divides itself into a few major subdivisions, we find that the path that the program takes in solving a problem traverses a tree. Conversely, any time we know exactly how to solve a problem, we can program it by a process of repeatedly subdividing our goals; again, running the program is a process of traversing a tree.

Part II of this talk discusses how to proceed when we *do not* know how to solve a problem. That is to say, we will discuss programs that think and learn.

II. AN ARTIFICIAL INTELLIGENCE APPROACH TO BUSINESS DATA PROCESSING

The traditional solution of the business data processing problem has been the 'human wave' hand-written cobol program. Economic considerations dictate that in these times of ever increasing hardware capability per dollar, a more automated solution be found. The authors present a solution in the form of a somewhat smart computer program, LINDA. Included are historical background and a discussion of some of the most important methodological issues confronted.

If I had it to do over again, I'd do it differently. How many times have we heard this or thought it? How much learning is the application of pattern previously followed, to the pattern we momentarily see ourselves in? This process seems to be what gives us as thinking beings, our computational power.

Although trees are sufficient for the comprehension of Weissman's algorithm for the differentiation of polynomials, the authors do not believe they suffice for the kinds of skill which humans display in planning and problem solving. In the sequel, the authors introduce LINDA, a computer program, whose organization is not naturally isomorphic to a tree.

The idea of not using the tree as our fundamental conceptual framework is not a new one. Its history can be traced back to the earliest serious attempts to describe neurological activity by mechanisms at the synaptic level.[1] Its use in computer programming had led to a breakthrough in our ability to model complex

[1] *We direct the reader to the work of Warren McCulloch. Many of his essential papers are available in a paperback collection, (McCulloch 1970).*

reasoning processes.[2]

The specific processes we ask LINDA to model are ones which the executive traditionally delegates to other human beings - to applications programmers and systems analysts. These humans are responsible for translating the user's verbal requests into computer code.

When an executive requests a programming job, he actually specifies very little information. He expects the systems analyst to draw on his knowledge and experience to fill in the gaps in these specifications. For example, the executive will never describe the data base to be used - he will, at most, mention it by name, and often he will assume that the context will make obvious which one is meant.

The systems analyst discharges his task by writing specifications to the programmer. These specifications are supposed to be so complete, and so accurate, that the programmer need know nothing about the problem domain.

This process, of acquiring the executive's problem request, is performed by LINDA's Problem Acquisition Phase.

Once the program specifications are complete, a long process of codification follows as more and more decisions are made and more information tracked down. The programmer replaces the name of a data base with tape serial numbers, and picks an internal file name. He converts the description of a data base format, into, typically, a COBOL or PL/1 structure. He interprets special instructions from the analyst, such as "use a binary table search", by recalling, from his programmer's training, how to do it; or by looking it up in the textbook. Eventually, he gets to

[2] *See Section 3 of the Bibliography. Especially accessible to the non-specialist are the preface and introduction to (Winograd 1971) and the first three chapters of (Sussman 1973). Although they masquerade as parts of Ph.D. theses, these chapters read like science fiction.*

the point where he has written a program. He then turns the program over to the COBOL or PL/1 compiler, to finish the job.

The jobs of programmer and compiler, being thus similar, are performed by LINDA's Compilation Phase.

> Look once again at the above description of a programmer at work. Each step in the refinement of specifications to machine code is nothing but the replacement of a symbol by the object it represents. In the "convert to PL/1" mode, the programmer evaluates "the commerical automobile data base", noting:
> I will code "open the data base" as
> "OPEN FILE(COMMAUTO) RECORD SEQUENTIAL INPUT;"
> and I will use COMMAUTO as the "data base". Or, if the programmer is a program written in LISP, it could do the above by executing the program:

```
(LAMBDA NIL
   (PROG2
      (SETQ  OPEN_COMMAND
             (OPEN  (FILE  COMMAUTO)
                RECORD          SEQUENTIAL
INPUT))
      (SET Q  READ_COMMAND
             (READ  (FILE  COMMAUTO)
                (INTO                   (EVAL
RECORD_LAYOUT)) )) ))
```

> This implicit use of the FUNARG property is the first step in an evolution towards a non-treelike (Bobrow and Wegbreit (1972) control structure.

LINDA's Compilation Phase, being essentially tree-like, is programmable using traditional programming techniques (e.g., "throw it in a hash table"), similar to those used in the differentiation problem discussed earlier. As her Problem Acquisition phase is not so simply connected, it will be our subject for the rest of this discussion.

"pick up a big red block."

--- (Winograd 1971)

Like Winograd with his SHRDLU, we feel that the best way to show you how our LINDA works, is to illustrate our presentation with a sample dialogue. In all of the following human - computer exchange, the voice of the computer is in CAPITAL LETTERS, and that of the human, in small letters.

1. TYPE OF JOB?

   report

   LINDA asks the questions in our system. Understanding English sentences has proven to be one of the most complex tasks A. I. has faced. We hope to avoid most of it by letting the computer take more of the lead, since it is the weaker party. To some extent, we try to impose her syntax, on the user's semantics.

2. LIKE A PREVIOUS JOB?

   no

   Terseness is valued by executives. Here, LINDA had hope to find a similar job in her long-term memory, to help her understand this one better.

3. INPUT FILES?

   commauto

   When the commercial automobile file was new, LINDA noticed the user was using a rather long word rather often. She suggested a nickname, and she now accepts COMMAUTO.

4. AND?

   none

   LINDA is equipped to write programs that use data from two or more sources.

5. ANYTHING YOU WANT ME TO LEAVE OUT?

   exposure > 1000

   The user can enter mathematical expressions, when in an appropriate context. Theoretically, exposure is the best linear predictor of loss that is directly measurable in the field.

6. AND?

   none

   One of the great advantages of LISP is that it enables us to use the same code here as we did to produce (4). Everything in LISP is a subroutine.

7. HOW SHOULD THE DATA BE BROKEN DOWN?

   state

   Actuaries impose a system of coordinates, not necessarily orthogonal, on their sample space. Their reports are always projections onto the product of a small number of axes. The axes chosen for a report, are the breakdowns. By analogy to geometrodynamics, we picture reports as tensors, operating on our breakdowns as 1-forms, and our data, as vectors. One of the most prominent of these breakdowns, for legal and historical reasons, is state.

8. AND?

   year

9. AND?

   none

   It is more convenient for LINDA to collect all the breakdowns now, while the executive is thinking about them. Top-down specification by the executive would appear to be a natural way that he and LINDA can compromise, on how to communicate.

10. TOTALS?

    standard

    This is the payoff for the work explained in (9). One question/answer pair has done the work for several breakdowns. Just as LINDA detected the frequent use of "commercial automobile", through surreptitious observation, pattern matching on her own control structure could make her notice a commonly invoked pattern. She asks the user to give it a name. Apparently, the same kind of totals were specified by this user on numerous occasions. Cf. (Sussman 1973), Ch. 10.

11. PLEASE ENTER THE RANGE OF YEARS.

    1969 1973

    In framing this question, we benefit from a little context-determined information. YEARS has the property ORDERED. The symbolism must account for the topology of the sample space somewhere. In this case, we have intervals over the countable discrete topology. A stronger condition than ORDERED is CONTINUOUS, a weaker is DISCRETE. In fact, DISCRETE is so weak, that every breakdown is (at least) DISCRETE. Note that the set of order properties is ORDERED.

12. FOR EACH STATE AND EACH YEAR, WHAT QUANTITIES

DO YOU WANT DISPLAYED?

earned premium

13. AND?

earned exposure

14. AND?

average premium

15. AND?

none

We treat each of these answers as if a single word, the name of a column. the only attempt to use any of the information implicit in the multi-word nature of these names, it an attempt to find "first-order" relationships between the words. In this case EARN IS_A PROGRAM, for which PREMIUM IS_A very common ARGUMENT. SOCRATES, the language in which LINDA is written, allows us to hard-wire the most common instances of this type of knowledge. As LINDA reorganizes her knowledge, she eventually gets to notice such commonly made references, and to propose creating these new concepts. We can take advantage here of our tendency to write "descriptive" names, like EARNED PREMIUM. Thus, a reasonable strategy for multi-word titles would be to create them as special objects. LINDA would quickly see what else is true about the data our user normally wants there. Here, if EARN is a wired-in function, the connection will be made very quickly. Cf. (Fahlman 1975). For a discussion of the development, of Open-Pivot grammars in children, see M. Braine, The Ontogeny of English Phrase Structure, in Language, 39(1), pp. 1-14.

16. YEARS: ACCIDENT, POLICY, OR OTHER?

accident

Earning is a procedure for defining a projection onto the time axis. In this case, every loss is assigned to the year it OCCURRED, and the premium is uniformly distributed over the period the policy is in effect. Thus, a policy written July 1, 1974, would earn half its premium in 1974 and half in 1975.

17. CALENDAR, FISCAL, OR OTHER?

    calendar

    Jan. 1 to Jan. 1.

18. PLEASE SUPPLY A FORMULA FOR AVERAGE PREMIUM.

    earned premium/earned exposure

19. IS THERE ANYTHING WE'VE LEFT OUT?

    no

    By this point, the only thing the executive is interested in is who he is having lunch with.

20. PROGRAM SPECIFICATIONS PREPARED.

    ESTIMATED COST ON LINE: $50.00.

    ESTIMATED TURNAROUND ON LINE: 10 MINUTES.

    SHALL I SUBMIT THE JOB?

    yes

21. TYPE OF JOB?

    LINDA stands ready to begin another job. She will inform the executive when this job has completed its run.

What is going on here? How does LINDA know what questions to ask? To anser these questions, we will introduce the concept of a *frame*.

> A *frame* is a data-structure for representing a stereotyped situation, like being in a certain kind of living room, or going to a child's birthday party. Attached to each frame are several kinds of information. Some of this information is about how to use the frame. Some is about what one can expect to happen next. Some is about what to do if these expectations are not confirmed.
>
> --- (Minsky 1974)

At the top level (right after "good morning" and "how's the family?"), LINDA is in the "what kind of job" frame. She expects one of a few possible responses, each corresponding to a possible transfer to a specific type of program frame. The "report program" frame divides its job (obtaining report specifications), into several sub-jobs. It asks for data selection criteria, for breakdowns, for field names and field definitions. It delegates each of these sub-jobs, to the special frame designed to handle it. The "breakdown frame", for example, obtains details on the grouping and subtotaling of data.

A frame, then, behaves much like a human problem solver, or like the LISP programs of part I, in that it "keeps in mind" only a few pieces of knowledge, at any one time.

Thus far, the description of LINDA's operation has varied little, from the "divide and conquer" approach described in Part I. However, more complex, and subtle, relationships between frames play a role in LINDA's activity.

Let us return to our "what kind of job" frame. Certain information, common to all request types, is gathered to this node - who are you, have I done a similar job in the past - all of which ties in certain other frames to the picture. It would be a mistake to think that you close the door to a wider universe when you enter a frame. You can still see into adjacent frames, and not only the one you have just come from. The context within which the frame occurs is reflected in pointers to "nearby" frames. A call for certain information, not available in this frame, can initiate a scan to other, nearby frames.

When LINDA enters a frame, she commonly attempts to find values for that frame's variables. For example, when in the EARNING frame, she desired two pieces of information: the endpoints for earning periods (CALENDER, FISCAL, OR OTHER?) and the rule for assigning data to years (ACCIDENT, POLICY, OR OTHER?). When she gets these, she has completed this frame satisfactorially. This frame need not, however, return control to its caller. Attempts to bind variables will institute searches into other frames. For example, LINDA never left the scope of the "type of job" frame, in this dialogue, until the very end.

Thus, LINDA spends most of her time nested knee deep in control structure. She can step outside of context, as it were, and look at what is going on. We mentioned this in a small way when we talked about nicknames and multi-word names. In both cases, she asked the parsing frame to do her a favor, i.e., "if you see anybody with a real long name, let me know about it". This kind of "peripheral vision" has been called a demon (Charniak 1972). The nickname and multi-word demons return control to the parser, with the world almost imperceptibly altered. The parser does not "consciously" call these demons. He does not make any allowance, in his actions, for the possible presence of a demon. It is of the nature of demons, that they do their work independent of the normal tree-walking operation of the program. These two demons, as it happens, are content to remain in the unconscious. There are, however, demons which can disrupt the main flow of control: demons which can wrench control right out of the immediate context.

For example, one such demon is the IDUNNO demon, which is triggered by any completely unexpected answer. The first thing that he does is to see if he was just called during the previous user interchange -- two "question marks" in a row indicate that a user is having a problem of a higher logical type.

The significance of this type of behavior is to be found in Bateson's metaphor of the logical categories of learning. (Bateson 1972, pp. 279-302) That ability, to bring her own actions into consideration, allows LINDA to begin to show Level II behavior. Bateson's ideas may be summarized as follows (M. Levin, On Bateson's Logical Level of Learning Theory, MIT MAC TM 57, 1975):

> Zero learning or non-learning is a response to information which is the same each time the information is presented... It is somewhat questionable to call this "learning" at all, and therefore the term Zero Learning seems appropriate... The primary purpose of mentioning Zero Learning is to catalogue some situations such that if they were to change, this would be evidence for some sort of higher learning.

> Learning I is the category that includes most of the learning that is studied in animal experiments... learning I is about something other than learning itself. Its content is explicit enough to be readily described by an observer...
>
> It seems to be characteristic of Learning I that it has a certain objectivity in that if a lesson is not learned correctly there is abundant feedback to make the organism learn about and correct its mistake.
>
> Learning II is learning about the contexts in which Learning I occurs. It is learning how to perceive the world, how to form meaningful gestalts from the background of sensory information, what to pay attention to. Learning II is the process of forming habitual value judgements. It is the source of what psychologists call character. Learning II is learning to be a particular personality and therefore is not objective learning. The lessons of Learning II are not objective in the sense of being learned correctly or incorrectly and therefore not subject to simple revision by failure to coincide with external reality as with Learning I. ...Learning II is characterized by a high degree of stabilization compared to Learning I. Learning II is frequently self-reinforcing, and it is generally not corrected by the environment because of its subjective nature. It is Learning II that makes the world of human affairs interesting, as well as providing most of the problems of knots (to borrow a word from R. D. Laing). It accounts for diversity, and explains why different individuals follow such different paths in their development, instead of becoming pretty much homogenized. ..."

In developing in LINDA powers of generalization, we have forced her, on occasion, to consciously observe the context in which she acts. LINDA suggests a synonym for "commercial automobile" when she has noticed a user use the phrase several times, in the same way. She lets the user define the notion of "standard" totals, for "similar" reasons. In these examples, she has observed a pattern in her own history, and has applied it to the pattern she momentarily sees herself in. In acting upon her partial awareness of past events, LINDA has developed a memory she can learn from. The acquisition of this ability coincides with LINDA's departure from the realm of tree-like objects.

Not all of the self-organizing features just described are fully developed in the LINDA of our dialogue. The world of insurance program generation does not demand a great deal of sophistication in the directions of adaptibility and originality. After all, there is no Time, Space, or Physics to deal with, let alone politics and religion. LINDA needs her abilities only because she must communicate, as naturally as possible, with a human being. Once specifications are complete, LINDA's job degenerates to that of today's compiler. Any ability LINDA will have to understand her employers will be a vast improvement over today's software.

As regards our own plans for LINDA, we hope that her initial development will give her the conceptual tools she needs for the gradual expansion of her high level functions. We want to give LINDA a certain set of goals or purposes, as well as a certain ability to reason and write programs. She is asked to minimize the burden on the user, both by keeping context from becoming too confusing for him, and by encouraging short and simple responses. She can best minimize her demands on the executive, by becoming more aware of the context in which he acts: That is, by coming to understand him better. LINDA already has the rudiments of pivot-word grammar. It is not inconceivable that she might someday progress to an understanding of natural human language. At such a point, we would indeed have a computer program that could think and learn.

APPENDIX: GRAPHS AND TREES

A graph is a simplicial complex of dimension 1. A tree is an arcwise connected graph, such that, for each 1-simplex element, s, of T, [T] - (s) is not connected. Alternatively, a tree is a simply connected graph. It can be shown that every tree is contractable (i.e. homotopic) to a point. It turns out that the interesting characteristic of different graphs can be expressed by their Euler characteristic. The Euler characteristic, E, is equal to the number of vertices minus the number of 1-simplices, E = 1, for all trees.

It can be shown that the following is true: let k be an arcwise connected graph. Let n be the maximum number of open 1-simplices that can be removed from k without disconnecting the space (this is the number of "basic circuits" in the network). Then n = 1 - E.

If v_0 is the vertex of a graph K, then $\pi(K,v_0)$ is isomorphic to the edge group, which is the set of equivalent classes of routes beginning and ending with v_0. This is, furthermore, isomorphic to the free group on n = 1 - E generators.

We therefore believe that the graph representation will give us a strong feeling for the complexity of the network, and that this approach may yield information on how to construct such networks. (cf. Fahlman 1975).

REFERENCES

Conceptual Roots

McCulloch, W. S., "Embodiments of Mind," MIT, 1970.
Wiener, N., "Cybernetics," MIT, 1961, 2nd Edition.
Minsky, M. and Papert, S., "Perceptrons," MIT, 1972, 2nd Edition.
Bateson, G., "Steps to an Ecology of Mind," Ballentine, 1972.

Logic

Shoenfield, J. R., "Mathematical Logic," Addison Wesley, 1967.
Levin, M., "Mathematical Logic for Computer Scientists," MIT Project MAC, MAC TR 131, June 1974.
Polya, G., "Patterns of Plausible Inference," Princeton, 1968, 2nd Edition.

Artificial Intelligence

Feigenbaum, E. A. and Feldman, J., ed., "Computers and Thought," McGraw-Hill, 1963.

Minsky, M., ed., "Semantic Information Processing," MIT, 1968.

Winston, P. H., "Learning Structural Descriptions from Examples," MIT Artificial Intelligence Laboratory, AI TR 231, Sept., 1970.

Winograd, T., "Procedures as a Representation for Data in a Computer Program for Understanding Natural Language," MIT Artificial Intelligence Laboratory, AI TR 235, Feb., 1971.

Charniak, E., "Toward a Model of Children's Story Comprehension," MIT Artificial Intelligence Laboratory, AI TR 266, Dec., 1972.

Sussman, G. J., "A Computational Model of Skill Acquisition," MIT Artificial Intelligence Laboratory, AI TR 297, Aug., 1973.

McDermott, D. V., "Assimilation of New Information by a Natural Language-Understanding System," MIT Artificial Intelligence Laboratory, AI TR 291, February 1974.

Minsky, M., "A Framework for Representing Knowledge," MIT Artificial Intelligence Laboratory, AI Memo 306, June 1974.

Fahlman, S. E., "A System for Representing and Using Real-World Knowledge," MIT Artificial Intelligence Laboratory, AI Memo 331, May 1975.

Languages for Artificial Intelligence

McCarthy, J. et al., "LISP 1.5 Programmer's Manual," MIT, 1965.

Weissman, C., "LISP 1.5 Primer," Dickenson, 1967.

Bobrow, D. and Weizenbaum, J., "List Processing and Extension of Language Facility by Embedding," IEEE Trans. Electronic Computers, *13*, No. 8, pp. 395-400, 1964.

Hewitt, C., "Description and Theoretical Analysis (Using Schemata) of Planner: A Language for Proving Theorems and Manipulating Models in a Robot," MIT Artificial Intelligence Laboratory, AI TR 251, April 1972.

Sussman, G. J., Winograd, T. and Charniak, E., "MICRO PLANNER Reference Manual," MIT Artificial Intelligence Laboratory, AI Memo 203, July 1970.

McDermott, D. V. and Sussman, G. J., "The Conniver Reference Manual," MIT Artificial Intelligence Laboratory, AI Memo 259, May 1972.

Bobrow, D. and Wegbreit, B., "A Model and Stack Implementation of Multiple Environments," CACM, *16*, No. 10, p. 39, October 1973. Also BB&N Report #2334, 1972.

Computer Programming

Knuth, E. D., "The Art of Computer Programming," Addison-Wesley. Of the seven volumes, three have been published to date.

APL FOR ACTUARIES

Donald L. Orth

IBM Corporation
1700 Market Street
Philadelphia, Pennsylvania

The purpose of this paper is to illustrate the use of APL as a mathematical notation. Many readers will be familiar with some fundamental advantages of APL as a programming language; the interactive approach to computing; the simple forms of input-output; the concept of programs as functions; the concept of arrays as the basic objects to which functions apply; the set of simple and yet powerful primary functions for construction and rearrangement of arrays and selection of subarrays. All of these matters concern the definition of algorithms in the language, and the suitability of a language for precise definitions of algorithms is one measure of its value as mathematical notation. Mathematics is also concerned with *analysis*, that is, with the study of relationships between quantities, and another measure of a language as a mathematical notation is how well it serves as a vehicle for studying these relationships. This paper is concerned with APL as both a vehicle for defining algorithms and a tool of analysis.

The APL primary functions, that is, the functions for which APL provides symbols, apply to arrays in various ways. It often

ISBN 0-12-394680-8

happens in defining new functions, called *secondary* functions, that we do not define them in such a way as to extend to arrays in ways similar to those of the primary functions. For instance, we may define a function that applies to vectors, but not to vectors along an axis of a higher rank array. Then later when we find, for example, that it would be convenient to apply the function to the rows of columns of a matrix, we must redefine the function. Had we taken the care to analyze the function and generalize its definition to apply to arrays, subsequent alterations in the definition would not be necessary. Also, once the function applies properly to arrays it can be viewed as a new primary function, and we can think of the set primary functions as being extended into our particular area of interest.

There are practical reasons for analyzing defined functions. First of all, it is rare that the analysis will not teach us something about the function or its applications. Secondly, the result of analysis is often not only a more generally applicable function but one that is defined in terms of just a few primary functions. In these cases it may be possible to recognize a relationship between the function and one produced for a seemingly unrelated application. These relationships often suggest new relations within each applications area which might otherwise remain hidden. In effect, we are using the primary function in a way similar to the way mathematicians use axiom systems, the difference being that the structures we are describing are new functions.

Two examples of the analysis of secondary functions, which deal with compound interest, are presented below. In both cases we begin with obvious, but restricted, function definitions and proceed to more general definitions. The examples serve as checks on our analysis at each step. The analyses use formal mathematical results, including identities phrased in APL notation, as well as informal computational experiments conducted at a terminal. Identities, which signify the equivalence of two expressions, are written with one of two expressions directly below the other. Labels of identities and expressions appear near the right margin, with the reference number enclosed in square brackets. Readers unfamiliar with APL may wish to consult Iverson's *An Introduction to APL for Scientists and Engineers*, publication no. 320-3019, IBM Corporation.

As our first example, consider a scalar D, which represents an initial deposit in a bank account, and a vector R, whose Ith element R[I] represents the rate of interest in effect on the Ith anniversary of the initial deposit. The problem is to determine the vector of balances B, whose Ith element B[I] represents the principal on deposit on the Ith anniversary, assuming no new deposits have been made. That is,

```
    B[1]=D                                          [1]
    B[I+1]=B[I]×1+R[I]                              [2]
```

Vectors of balances are produced by the following function.

```
    ∇ B←R BL D;I
[1]   B←,D
[2]   I←0
[3]  TEST:→((ρR)<I←I+1)/0
[4]   B←B,B[I]×1+R[I]
[5]   →TEST
    ∇
```

For example:

```
    D←100
    R←0.06 0.07 0.055
    R BL D
100 106 113.42 119.6581
```

The function BL produces vectors of balances element-by-element. These vectors can also be produced in closed form. To do this we will use an extension of the inverse relations between times and divide. The function PQ:

```
    ∇ Q←PQ V
[1]   Q←(1↓V)÷¯1↓V
    ∇
```

produces vectors of "pairwise quotients" of its argument vectors. For example:

```
      V←1 5 15 150
      PQ V
5 3 10
      V[2 3 4]÷V[1 2 3]
5 3 10
```

The function called ATS (for augmented times scan) is defined as follows:

```
      ∇ W←U ATS V
[1]     W←×/U,V
      ∇
```

For the functions PQ and ATS, and for scalars S and vectors V, we have the identities:

```
      PQ S ATS V                    V[1] ATS PQ V              [3]
      V                             V
```

For example:

```
      V
1 5 15 150
      PQ 6 ATS V
1 5 15 150
      V[1] ATS PQ V
1 5 15 150
```

The element-by-element equations 2 can be written as

```
      (B[I+1]÷B[I])=1+R[I]
```

and evidently these equations are equivalent to the equation

```
      ∧/(QT B)=1+R                                              [4]
```

According to the right hand side of 3, equations 4 and 1 are equivalent to

```
      ∧/B=D ATS 1+R
```

Thus for scalar right arguments and vectors left arguments, the function BL is equivalent to the following function.

```
      ∇ B←R BLA D
[1]     B←D ATS 1+R
      ∇
```

For example:

```
       R
0.06 0.07 0.055
       D
100
       R BL D
100 106 113.42 119.6581
       R BLA D
100 106 113.42 119.6581
```

The letter A in the name BLA suggests that this function applies to array arguments. For example, consider the vector

```
       D
100 150 235
```

and the matrix

```
        R
      0.08        0.07        0.07        0.06
      0.06        0.08        0.09        0.06
      0.08        0.07        0.09        0.08
```

Each element of D is to be thought of as an initial deposit, and each row of R is to be thought of as a vector of interest rates. The result R BLA D is a matrix whose Jth row, for each index J, is the vector of balances for the initial deposit D[J] and the vector of interest rates R[J;]. The two-place approximation to this result is given by:

```
       2⍕R BLA D
100.00 108.00 115.56 123.65 131.07
150.00 159.00 171.72 187.17 198.41
235.00 253.80 271.57 296.01 319.69

       2⍕R[1;] BL D[1]
100.00 108.00 115.56 123.65 131.07
       2⍕R[2;] BL D[2]
150.00 159.00 171.72 187.17 198.41
       2⍕R[2;] BL D[3]
235.00 253.80 271.57 296.01 319.69
```

Consequently, the function BLA applies to arrays in much the same way as primary functions.

Our second example is an extension of the previous compound interest problem, where we allow for new deposits to be made on the anniversaries of the initial deposit. Precisely, consider a vector R of interest rates and a vector D, whose first element is the initial deposit and whose element D[[I+1] is the deposit to be made on the Ith anniversary of the initial deposit. Evidently, (ρD)=1+ R. As before, the problem is to determine the vector B whose element B[I] represents the principal on deposit on the Ith anniversary. That is,

```
    B[1]=D[1]
    B[I+1]=D[I+1]+B[I]×1+R[I]
```

Vectors of balances are produced by the following function.

```
    ∇ B←R BAL D;I
[1]    B←1↑D
[2]    I←0
[3]   TEST:→((ρR)<I←I+1)/0
[4]    B←B,D[I+1]+B[I]×1+R[I]
[5]    →TEST
    ∇
```

For example:

```
    R←0.05 0.08 0.06
    D←120 50 40.50 75

    R BAL D
120 176 230.58 319.4149
```

In order to analyze the function BAL we will consider it as a function of its left argument and as a function of its right argument separately. We begin with the right argument. Specifically, we will study, for a specified vector R, the function called RT (for <u>r</u>igh<u>t</u> argument) defined as follows:

```
    ∇ B←RT D
[1]   B←R BAL D
    ∇
```

In order to produce examples during analysis we will specify a vector of interest rates, say

```
     R←0.06 0.05 0.07
```

The analysis of RT proceeds by a straightforward application of the principles of linear algebra, and is based on the fact that RT is a linear function. That is, for vectors X and Y and scalars S:

```
     RT X+S×Y
     (RT X)+S×RT Y
```

For example:

```
      X←100 50 23 48
      Y←34 57 111 79
      S←6
      RT X+S×Y
304 714.24 1438.952 2061.67864
      (RT X)+S×RT Y
304 714.24 1438.952 2061.67864
```

Since RT is a linear function it can be represented as a +.× inner product. That is, there is a matrix M, depending on the vector of interest rates R but not on a vector of deposits D, for which:

```
     RT D                                                    [5]
     D+.×M
```

The matrix M is produced by applying the function RT to a canonical basis of the vector space of dimension ⍴D (or equivalently, 1+⍴R). Precisely, the function called ID:

```
     ∇ I←ID N
[1]    I←(⍳N)∘.=⍳N
     ∇
```

produces identity matrices, whose rows are elements of canonical bases. For example:

```
     ID 3
1 0 0
0 1 0
0 0 1
```

The row M[J;] of the matrix M described in identity 5 is identical to RT (ID 1+ρR)[J;]. Thus the following function, called MRT, produces the matrices in the +.× representation of RT.

```
     ∇ M←MRT I;J
[1]   M←(0,¯1↑ρI)ρ0
[2]   J←1
[3]   AGAIN:M←M,[1] RT I[J;]
[4]   →((1↑ρI)≥J←J+1)/AGAIN
     ∇
```

For example, for the vector R specified above:

```
      R
0.06 0.05 0.07
      M←MRT ID 1+ρR
      M
     1          .05      1.134      1.21338
     0         1         1.08       1.1556
     0         0         1          1.07
     0         0         0          1
      D←45 68 103 27
      D
      RT D
45 115.7 224.485 267.19895
      D+.×M
45 115.7 224.485 267.19895
```

An examination of the functions RT and MRT shows that they are functions of the vector of interest rates R, even though R does not appear as an explicit argument in either function. This deficiency is overcome in the function called RM (for rate matrix), defined as follows:

```
     ∇ M←RM R;I;J
[1]    I←ID 1+ρR
[2]    M←(0,¯1↑ρI)ρ0
[3]    J←1
[4]    AGAIN:M←M,[1] R BAL I[J;]
[5]     →((1↑ρI)≥J←J+1)/AGAIN
     ∇
```

Continuing the preceding example:

```
      ∧/,(MRT ID 1+ρR)=RM R
1
```

To summarize, for a vector of interest rates R and a vector of deposits D:

```
     R BAL D                                      [6]
     D+.×RM R
```

Using the second expression in identity 6 as a guide, we define the function called BALA as follows:

```
     ∇ B←R BALA D
[1]   B←D+.×RM R
     ∇
```

The function BALA applies to arrays D and vectors R. Specifically, if D is a matrix whose row D[J;], for each index J, represents a vector of deposits, then the result R BALA D is a matrix whose Jth row is the vector of balances for the vector of interest rates R and the vector of deposits D[J;]. For example:

```
     R
0.06 0.055 0.08

     D
 23 45 123 89
110 34  81 29
 95 74 134 45

     2⍕R BALA D
 23.00  69.38 196.20 300.89
110.00 150.60 239.88 279.07
 95.00 174.70 318.31 388.77

     2⍕R BAL D[1;]
23.00 69.38 196.20 300.89

     2⍕R BAL D[2;]
110.00 150.60 239.88 279.07

     2⍕R BAL D[3;]
95.00 174.70 318.31 388.77
```

This completes the analysis of BAL as a function of its right argument.

In order to analyze BAL as a function of its left argument, it is enough to analyze the function RM. For that task we have some guidance from our first example. Namely, the previously defined function BL can be defined in terms of the vector ×/1,1+R, and elements of the latter vector are times reductions of subsets of the vector 1+R. Thus it is reasonable to test whether or not elements of the matrix RM R can also be defined in terms of times reductions of subsets of 1+R. Before we test this hypothesis, however, we will examine the relation between times reduction and ×.* inner products.

Subsets of a vector V can be represented by a logical vectors of length ρV, where 1 occurs in position J of a logical vector if and only if the Jth element of V is a member of the corresponding subset. For example, the vector 1 1 0 1 represents the subset 1 2 4 of the vector 1 2 3 4. Moreover, if the logical vector L represents a subset of a vector V, then ×/L/V is the product of the elements of that subset. For example:

```
      V←1 2 3 4
      L←1 1 0 1
      L/V
1 2 4
      ×/L/V
8
```

Also, element J of the vector V*L is 1 if L[J] is 0 and V[J] if L[J] is 1. Thus ×/V*L, or equivalently V×.*L, is equivalent to ×/L/V. For example:

```
      V
1 2 3 4
      L
1 1 0 1
      ×/L/V
8
      V×.*L
8
```

In general, for arrays A and L, where L is a logical array, the elements of the array A×.*L are times reductions of subsets of the vectors along the last axis of A, as represented by logical vectors along the first axis of L. For example:

```
      A←4 3ρι12
      A
 1  2  3
 4  5  6
 7  8  9
10 11 12
      L←3 2ρ1 0 1 1 0 1
      L
1 0
1 1
0 1
      A×.*L
  2                6
 20               30
 56               72
110              132
      (A×.*L)[1;2]
6
      (A×.*L)[4;1]
110
      ×/L[;2]/A[1;]
6
      ×/L[;1]/A[4;]
110
```

We will now return to the conjecture that the elements of a matrix RM R are times reductions of subsets of the vector 1+R, and test the conjecture with an example.

```
      R
0.05 0.08 0.07
      M←RM R
      M
   1          1.05       1.134      1.21338
   0          1          1.08       1.1556
   0          0          1          1.07
   0          0          0          1
```

Evidently our conjecture must be modified, for the zero elements in the matrix M cannot be produced in the desired manner unless the vector 1+R has at least one zero element. However, if we work with vector 0,1+R in place of 1+R, it will be possible to produce the zero elements. We will proceed through the matrix element-by-element in row major order, starting with M[1;1].

The logical vectors representing the appropriate subsets of R will be accumulated as rows of a matrix called P. The element M[1;1] is 1, or equivalently (0,1+R)×.*0 0 0 0. Define the matrix P as follows:

```
      P←1 4ρ0 0 0 0
```

Element M[1;2] is 1+R[1], or equivalently (0,1+R)×.*0 1 0 0. Append the vector 0 1 0 0 to the bottom of P as follows:

```
      P←P,[1] 0 1 0 0
      P
0 0 0 0
0 1 0 0
```

Element M[1;3] is (0,1+R)×.*0 1 1 0

```
      M[1;3]
1.134
      (0,1+R)×.*0 1 1 0
1.134
      P←P,[1] 0 1 1 0
```

Element M[1;4] is (0,1+3)×.*0 1 1 1

```
      M[1;4]
1.21388
      (0,1+R)×.*0 1 1 1
1.21338
      P←P,[1] 0 1 1 1
```

Element M[2;1] is (0,1+R)×.*1 0 0 0

```
      M[2;1]
0
      (0,1+R)×.*1 0 0 0
0
      P←P,[1] 1 0 0 0
```

Proceeding through the matrix M in this manner results in the following matrix P.

```
      P
0 0 0 0
0 1 0 0
0 1 1 0
0 1 1 1
1 0 0 0
0 0 0 0
0 0 1 0
0 0 1 1
1 0 0 0
1 0 0 0
0 0 0 0
0 0 0 1
1 0 0 0
1 0 0 0
1 0 0 0
0 0 0 0
```

Now form the array A←4 4 4ρ⍉P. Then M is (0,1+R)×.*A.

```
      A←4 4 4ρ⍉P
      A
0 0 0 0
1 0 0 0
1 1 0 0
1 1 1 0

0 1 1 1
0 0 0 0
0 0 0 0
0 0 0 0

0 0 1 1
0 0 1 1
0 0 0 0
0 0 0 0

0 0 0 1
0 0 0 1
0 0 0 1
0 0 0 0
      ∧/,M=(0,1+R)×.*A
1
```

The relatively simple pattern in the array A can be produced as follows:

```
      I←⍳4
      (I∘.>I),[1] ∨/(1↓I)∘.=(0.5+I)∘.⌈I
0 0 0 0
1 0 0 0
1 1 0 0
1 1 1 0

0 1 1 1
0 0 0 0
0 0 0 0
0 0 0 0

0 0 1 1
0 0 1 1
0 0 0 0
0 0 0 0

0 0 0 1
0 0 0 1
0 0 0 1
0 0 0 0
```

This suggests the following definition of the function called ARRAY:

```
      ∇ A←ARRAY N;I
[1]     A←(I∘.>I),[1]∨/(1↓I)∘.=(0.5+I)∘.⌈I←⍳N
      ∇
```

For the vector R in the above example we see that the matrix RM R is identical to (0,1+R)×.*ARRAY 1+⍴R. In fact, this is true for every vector R. That is:

```
      RM R
      (0,1+R)×.*ARRAY 1+⍴R
```

Combining this identity with identity 6 yields the following identity for vectors R:

```
      R BAL D                                        [7]
      D+.×(0,1+R)×.*ARRAY 1+⍴R
```

If R is an array of rank greater than 1, then the function of R defined by the expression (0,1+R)×.*ARRAY 1+¯1↑⍴R applies to vectors along the last axis of R, and matrices along the last two axes of the results are identical to values of RM when applied to the vectors along the last axis of R. For example:

```
      R
   0.7          0.9          0.8          0.9
   0.5          0.6          0.8          0.5
   0.5          0.6          0.5          0.7
      B←(0,1+R)×.*ARRAY 1+¯1↑ρR
      ∧/,B[1;;]=RM R[1;]
1
      ∧/,B[2;;]=RM R[2;]
1
      ∧/,B[3;;]=RM R[3;]
1
```

The next-to-last axis of B can be transposed to the first axis as follows:

```
      C←(⍋⌽(ρρR)=⍳ρρB)⍉B
```

Then if D is a vector of deposits, the rows of the matrix D+.×C are vectors of balances for the vectors of interest rates represented by the rows of R.

```
      D
79 46 102 94 56
      2⍕D+.×C
79.00 180.30 444.57 894.23 1755.03
79.00 164.50 365.20 751.36 1183.04
79.00 164.50 365.20 641.80 1147.06
      2⍕R[1;] BAL D
79.00 180.30 444.57 894.23 1755.03
      2⍕R[2;] BAL D
79.00 164.50 365.20 751.36 1183.04
      2⍕R[3;] BAL D
79.00 164.50 365.20 641.80 1147.06
```

The above example suggests the following definition of the function called RS:

```
      ∇ S←RS R
[1]     S←(0,1+R)×.*ARRAY 1+¯1↑ρR
[2]     S←(⍋⌽(ρρR)=⍳ρρS)⍉S
      ∇
```

For vectors of interest rates R and vectors of deposits D identity 7 becomes:

```
      R BAL D                                              [8]
      D+.×RS R
```

The second expression in identity 8 is the basis of the function called BALANCE, which is defined as follows:

```
     ∇ B←R BALANCE D
[1]    B←D+.×RS R
     ∇
```

The function BALANCE applies to arrays R and D of arbitrary rank for which (¯1↑⍴D)=1+1↑⍴R, and vectors along the last axis of an array R BALANCE D are vectors of balances for vectors of interest rates represented by vectors along the last axis of R and vectors of deposits represented by vectors along the last axis of D. For example:

```
      R
    0.7          0.9          0.8          0.9
    0.5          0.6          0.8          0.5
    0.5          0.6          0.5          0.7

      D
60 130 150  80  70
90 140 120  70 120
      AA←R BALANCE D
      2⍕AA[1;2;]
60.00 220.00 502.00 983.60 1545.40
      2⍕R[2;] BAL D[1;]
60.00 220.00 502.00 983.60 1545.40
```

Consequently, just as with the function BLA, the function BALANCE applies to arrays in a way similar to the primary functions.

CONCLUDING COMMENTS

The following observations occurred to the author while working out the preceding examples. These observations, as well as the analysis in the examples, may suggest other questions to readers familiar with variations of interest rate problems.

1. The representation of compound interest in terms of the times scan function must be a generalization of the more familiar

representation in terms of the power function. This suggest the identity:

```
×\1,1+R
(1+R[1])*0,ιρR
```

for vectors R, all of whose elements are equal.

2. Readers familiar with APL may have noticed the relation between the function BAL and the primary base value function. That is:

```
¯1↑R BAL D
(1,1+R)⊥D
```

This identity could be used in producing more efficient definitions of the functions BAL and BALANCE.

3. The final observation concerns the computational efficiency of the function RS. Specifically, it is not necessary to evaluate the function ARRAY each time RS is evaluated. Suppose we know that we will always work with vectors of deposits of length no more than 30. Define A̲←ARRAY 30 and then define the function called ARRAY̲ as follows:

```
    ∇ A←ARRAY̲ N
[1]   A←(N,N,N)↑A̲
    ∇
```

Then ARRAY N and ARRAY̲ N are identical for integers N less than or equal to 30.

BACKWARD POPULATION PROJECTION BY A GENERALIZED INVERSE

T. N. E. Greville

National Center for Health Statistics
Rockville, Maryland

Nathan Keyfitz

Department of Sociology
Harvard University
Cambridge, Massachusetts

The Leslie population projection matrix may be used to project forward in time the age distribution or age-sex distribution of a population. As it is a singular matrix, it does not have an inverse, and so it is not clear that there is a corresponding procedure for backward projection. In terms of the eigenvalues and eigenvectors of the Leslie matrix, certain generalized inverses are constructed that can sometimes be used advantageously for backward projection.

ISBN 0-12-394680-8

1. THE PROBLEM

Projection is the technique by which demographers work out future populations, assuming fixed or changing age-specific rates of birth, death, and migration. A principal way in which one can interpret present conditions and trends is to see what would happen if they were continued into the future.

A simple form of projection starts with the age-sex distribution of a population, say as obtained from a recent census, and applies to it a fixed set of age-sex-specific rates of death and of births by age of mother. For simplicity migration is excluded. The trajectory as the population moves hypothetically through time is worked out for each age and sex group. The arithmetic was originally carried out for successive time intervals, without any formal mathematics, by Cannan (1895), Bowley (1924), and Whelpton (1936). Later Lewis (1942), Bernardelli (1941), and Leslie (1945) noted that the arithmetic constituted a linear process, and hence it could be arranged in matrix form. Though the matrix offers no advantage for computing, it provides an understanding of projection, and demography owes much to Leslie, who worked out its properties in considerable detail.

The awkwardness of backward projection, for example for a time prior to the beginning of a series of censuses, contrasts with the straightforwardness of forward projection. Once the age-specific rates of birth and death of a population are assumed there is no difficulty in saying where it is going, and theorems exist showing that it ultimately attains a stable condition. In fact, continued forward projection by the Leslie matrix can be shown to lead to the stable age distribution provided only there are two fertile ages prime to one another. Like any ergodic process, this one may be described as forgetting its past: irrespective of the original age distribution, the same stable age distributions will be reached.

That the identical stable condition is reachable through different paths is what makes impossible the retrieval of the path once the stable condition has been reached. If a plane is covered with lines, defined for example by a differential equation, one can in general proceed in either direction from a point, following the unique line through that point. But where a number of paths converge in a focus or other singularity, the trail is lost; a path cannot be followed through a singularity. Our stable population constitutes such a singularity in a space that has one dimension for each age category, and the problem is to trace a path back from the singularity or from a point near it.

It seems a paradox of the pure theory of population dynamics that the past is less easily accessible than the future. The difficulty arises because the Leslie matrix is singular. Methods for backward projection can be devised that correspond to the several reasonable ways of calculating a generalized inverse of a singular matrix. If we wish to go a short distance back we may preserve in the inverse several of the original eigenvalues; if we wish to go a longer way back, or if we are already very close to stability, we must settle for only the dominant eigenvalue and give up hope of capturing the waves through which the population reaches its resting point.

2. A GENERALIZED INVERSE OF THE LESLIE MATRIX

Let the Leslie matrix A have n rows and n columns, and be of rank n - 1. Its nonzero elements are confined to (i) the subdiagonal, and (ii) the top row. The subdiagonal elements are survival rates from a given age interval to the next age interval. The elements of the top row are essentially fertility rates, and so they vanish after the first m, where $m < n$ (Leslie, 1945).

A generalized inverse of a given singular matrix must correspond in some way to the given matrix and have some of the properties of the classical inverse of a nonsingular matrix. Usually, there are many generalized inverses, and the choice of one from among them depends on the use to be made of it. If A is the given matrix, it is usual to choose a generalized inverse X that satisfies at least one, and preferably both of the two relations (Rao and Mitra, 1971)

$$AXA = A \tag{1}$$

and

$$XAX = X. \tag{2}$$

The Leslie matrix transforms the age distribution of a population at time t into its age distribution at time $t + \Delta t$. The first attempt at a generalized inverse might be a superdiagonal matrix of order n, which we shall denote by X_0, whose superdiagonal elements are the reciprocals of the corresponding subdiagonal elements of the Leslie matrix A. This matrix X_0 transforms the later age distribution back into the earlier one, except for the population of the final age interval, which was annihilated by the forward transformation.

In fact, X_0 is a reasonably satisfactory generalized inverse of A. Like A, it is of order n and rank $n - 1$. It is easily verified that it satisfies both (1) and (2). A disadvantage of X_0, however, lies in the fact that by its use the population in the oldest age interval always comes out zero.

This is easily remedied by observing that properties (1) and (2) are retained if the 0's in the bottom row of X_0, except the first, are replaced by arbitrary elements. This amounts to estimating the number in the final age interval at time t as some linear combination of the numbers at time $t + \Delta t$ in all age intervals except the youngest.

It is clear that there are many ways of arriving at such a linear estimate. One approach involves the eigenvalues and eigenvectors of the two matrices. It is an obvious property of the classical inverse of a nonsingular matrix H, that if

$$Hx = \lambda x,$$

then

$$H^{-1}x = \lambda^{-1}x.$$

In words, a nonsingular matrix and its inverse have common eigenvectors, associated with respective eigenvalues that are reciprocals of each other. It has been shown (Greville, 1968) that it is possible for a singular matrix and a generalized inverse to share somewhat similar properties.

Leslie (1945) suggested that three eigenvalues of the matrix are significant: (i) the real "Perron root", which we shall denote by λ_1, and (ii) a conjugate pair of complex eigenvalues, which are those closest to the Perron root in absolute value, which we denote by λ_2 and $\overline{\lambda}_2$. The Perron root is related to the rate of natural increase of the population that would result if the mortality and natality conditions reflected in the Leslie matrix were perpetuated, while the pair of complex roots is related to the amplitude and period of the oscillations which, under the hypothesis, would precede the attainment of a stable state.

Evidently there is a real cubic polynomial with leading coefficient unity whose three zeros are precisely the reciprocals of these three eigenvalues. Let this polynomial be

$$q(z) = (z-\lambda_1^{-1})(z-\lambda_2^{-1})(z-\overline{\lambda}_2^{-1}) = z^3 + c_2z^2 + c_1z + c_0, \qquad (3)$$

and let p_i denote the survival rate from the ith age interval to the (i + 1)th (which is the ith subdiagonal element of A). Then, if we take X_1 to be a matrix like X_0 except that the last three elements of the bottom row are

$$-c_0p_{n-1}p_{n-2}, \ -c_1p_{n-1}, \ -c_2 \qquad (4)$$

(instead of 0's), it is easily verified that the characteristic polynomial of X_1 is $z^{n-3}q(z)$. Thus, the eigenvalues of X_1

consist of n - 3 0's and the reciprocals of λ_1, λ_2 and $\overline{\lambda}_2$. A and X_1 have in common the eigenvectors associated with these three eigenvalues.

The use of X_1 in attempting to reverse the transition effected by A amounts to estimating the population at time t in the final age interval on the basis of the rate of natural increase implied by A and the period and amplitude of the principal oscillations that would precede attainment of the stable state under the operation of A.

The generalized inverse X_1 has, however, a possible disadvantage. It is easily seen that c_0 in (3) must be negative, because $-c_0$ is the product of the three zeros of q(t), which can be expressed as the product of two positive quantities, viz. the reciprocal of the Perron root, and the product of two complex conjugates. This is all to the good, but we now find that the last two elements (4) cannot both be nonnegative. This can be shown as follows. If μ is the reciprocal of the Perron root and $\alpha + \beta_i$ and $\alpha - \beta_i$ are the remaining zeros of q(t), then, by the definition of the Perron root, the latter two zeros must have absolute value greater than μ. That is

$$\alpha^2 + \beta^2 > \mu^2.$$

Now, since c_2 is the negative of the sum of the zeros of q(t) and c_1 is the sum of the pairwise products of the zeros, we have

$$c_1 = \alpha^2 + \beta^2 + 2\alpha\mu, \qquad c_2 = -2\alpha - \mu.$$

Thus, if $-c_2$ is nonnegative, this implies $\alpha > (1/2)\mu$, and therefore

$$c_1 > \mu^2 - \mu^2 = 0,$$

so that $-c_1$ is negative. This leaves open the theoretical possibility that use of X_1 to retroject the age distribution might produce a negative population in the oldest age interval at time t.

In such a case one would disregard the complex eigenvalues and take X_1 the same as X_0 except that the lower right-hand element of the entire matrix is $\mu = \lambda_1^{-1}$ instead of 0. This amounts to estimating the population at time t in the final age interval on the basis of the rate of natural increase alone.

Note that in theory one could go much farther by taking $q(z) = z^{n-1} + c_{n-2}z^{n-2} + \cdots + c_0$ as the monic polynomial whose zeros are the reciprocals of all the n - 1 nonzero eigenvalues of A, and the elements of the bottom row of X_1 as

$$0, -c_0p_{n-1}p_{n-2}\cdots p_2, -c_1p_{n-1}\cdots p_3, \ldots, -c_{n-3}p_{n-1}, -c_{n-2}.$$

Then the characteristic polynomial of X_1 is $zq(z)$. Note that it could be obtained, except for a constant multiplier, by merely reversing the order of the coefficients in the characteristic polynomial of A. Thus, the eigenvalues of X_1 are 0 and the reciprocals of all the nonzero eigenvalues of A. X_1 is therefore a full spectral inverse of A, except for the eigenvectors associated with the zero eigenvalue. We did not try this, but suspect that it would produce erratic results.

The rule seems to be that for long-term projection backwards one cannot improve on the dominant root. For backward projection over a short interval the first three roots often seem to help. In general the shorter the interval over which one projects backwards the more possible it is to preserve minor roots without finding erratically large and impossibly negative populations.

3. RELATION TO THE CLASSICAL INVERSE

In studying the possible generalized inverses of the Leslie matrix A, it is convenient to express A in the partitioned form

$$A = \begin{bmatrix} f^T & 0 \\ P & 0 \end{bmatrix} \qquad (1)$$

where

$$f^T = [f_1 \quad f_2 \cdots f_{n-1}],$$

and

$$P = \mathrm{diag}(p_1, p_2, \ldots, p_{n-1}).$$

(In general, some of the quantities f_j are zero.)

It can be shown that the most general matrix X satisfying (2.1) and (2.2) is of the form

$$X = \begin{bmatrix} 0 & P^{-1} \\ 0 & u^T \end{bmatrix} \left(I_n + \begin{bmatrix} 0 \\ z \end{bmatrix} [1 \quad -f^T P^{-1}] \right), \tag{2}$$

where u and z are arbitrary (n - 1)-vectors. However, the interesting case is that in which z is a zero vector, so that we have

$$X = \begin{bmatrix} 0 & P^{-1} \\ 0 & u^T \end{bmatrix}. \tag{3}$$

Note the close analogy of (3) to (1).

It has been argued that a generalized inverse of the Leslie matrix should have the property that if the Leslie matrix is used to project an age distribution forward in time and then the inverse is applied to bring it back, the original age distribution should be recovered, and similarly if the inverse under consideration is used first to project backward, and then the Leslie matrix to return to the starting point in time. It is clear, however, that these conditions cannot be fulfilled in general unless the matrix A is nonsingular and X is its classical inverse A^{-1}, so that

$$XA = AX = I.$$

Since the Leslie matrix is singular this is not the case, but it is of interest to see how close we can come to fulfillment of the conditions stated.

Even with X given by the most general expression (2), it is easily verified that

$$XA = \begin{bmatrix} I_{n-1} & 0 \\ u^T P & 0 \end{bmatrix},$$

and the original age distribution is, in fact, recovered, except for the oldest age interval, which is understandably lost.

In the reverse situation, the required condition is fulfilled if the original age-distribution vector, say u, was obtained by application of the matrix A from an earlier one, say y, because in this case,

$$AXu = AXAy = Ay = u.$$

Moreover, with X given by (3),

$$AX = \begin{bmatrix} 0 & f^T P^{-1} \\ 0 & I_{n-1} \end{bmatrix},$$

so that the original age distribution is recovered except for the youngest age interval, where fertility rates are involved.

4. APPLICATION

Let us test these suggestions by backward projection of the older U.S. female population from 1967; using data from that year only we will estimate for 1962. In all age intervals but the last this would be done by the reciprocal of the survival ratio, a procedure whose properties are straightforward and well-known.

From the vital statistics for 1967 the Perron root is $\lambda_1 = 1.0376$ and the roots following are $\lambda_2, \lambda_3 = 0.3098 \pm 0.7374i$. The reciprocals of these are $\mu = 0.9638$ and $\alpha \pm \beta i = 0.4843 \pm 1.1527i$. The polynomial q(z) of (2.3) is

$$q(z) = z^3 - z^2(1.9323) + z(2.4966) - 1.5065.$$

Probability of survival into the last age interval is $p_{n-1} = 0.8024$, and into the second-to-last is $p_{n-2} = 0.7030$. Hence, the bottom row of the inverse matrix ends up with the three numbers

$$\begin{bmatrix} \cdot & \cdot & \cdot & \cdot & \cdot & \cdot & \cdot & \cdot & \cdot \\ \ldots & 0.8498, & -2.0032, & 1.9323 \end{bmatrix}$$

and they are to premultiply the last three age intervals (75-79, 80-94, 85+) of the 1967 age distribution:

$$\begin{bmatrix} \cdots \\ 2198 \\ 1286 \\ 727 \end{bmatrix}$$

expressed in thousands. The inner product of the two triplets of numbers shown, constitutes the estimate on this particular way of forming the inverse, and it turns out to be 696 thousands. The observed 1962 figure was 602. (All numbers rounded from computer printout.)

On the procedure of projecting back by the reciprocal of the Perron root alone we obtain 727/1.0376 = 701, which is a slightly larger discrepancy.

For some other populations the superiority of the three roots method shows more clearly. For example using data on Belgium 1960 to estimate women aged 85 and over for 1955 the three roots method gives 31,323 and one root gives 33,543 against an observed 1955 figure of 27,880. Bulgaria 1965 projects back to 1960 on three roots at 17,550, on one root at 23,567, against an observed 15,995. In terms of percentage error, our three cases show:

	Three roots	One root
United States 1967	15.7%	16.4%
Belgium	12.3%	20.3%
Bulgaria 1965	9.7%	47.3%

But many other factors enter observations of the number of persons 85 years of age and over, including errors of enumeration, and one cannot count on the three roots method always being superior.

REFERENCES

Bernardelli, H. 1941. Population waves, *J. Burma Res. Soc. 31,* 1-18.

Bowley, A. L. 1924. Births and population of Great Britain, *J. Roy. Econ. Soc. 34,* 188-192.

Cannan, E. 1895. The probability of a cessation of the growth of population in England and Wales during the next century, *The Economic J. 5,* 505-515.

Goodman, L. A. 1968. An elementary approach to the population projection-matrix, to the population reproductive value, and to related topics in the mathematical theory of population growth, *Demography 5,* 382-409.

Greville, T. N. E. 1968. Some new generalized inverses with spectral properties, in "Theory and Application of Generalized Inverses of Matrices" (Texas Technological College Mathematics Series No. 4), pp. 26-46, Texas Tech. Press, Lubbock, Texas, 1968.

Leslie, P. H. 1945. On the use of matrices in certain population mathematics, *Biometrika 33,* 183-212.

Lewis, E. G. 1942. On the generation and growth of a population, *Sankhya 6,* 93-96.

Rao, C. R. and Mitra, S. K. 1971. "Generalized Inverse of Matrices and Its Applications," Wiley, New York.

Whelpton, P. K. 1936. An empirical method of calculating future population, *J. Amer. Stat. Assoc. 31,* 457-473.

ACCOUNTING PRINCIPLES FOR LIFE INSURANCE: REFLECTIONS ON LANGUAGE AND NOTATION

Gottfried Berger

Cologne Life Reinsurance Company
P. O. Box 300
Stamford, Connecticut 06904

INTRODUCTION

While trying to explain to Europeans actuarial developments in the U.S.A., particularly in connection with asset shares, the Anderson concept, GAAP, etc., I started some sort of a "Dictionary on Reserves".

My object was mainly to make a contribution to the discussion on reforming international actuarial notation. One point I intended to make was that a symbol such as ${}_tV_{x:n}$ incorporates

(1) indices t,x,n, which are rather unimportant; and

(2) the concept "V" of a reserve, which deserves further clarification.

In the course of this work I was led to meditate on the various languages we use for communicating and for arriving at logical conclusions, namely

(1) the "accountant's language", which depends heavily on carefully chosen, suggestive numerical examples;

ISBN 0-12-394680-8

(2) The "actuarial language", which is somewhere in the middle between (1) and

(3) classical algebra, or more generally, mathematics; and

(4) computer-oriented algebra, namely APL.

While trying to compare the powers and limitations of these languages by applying them to the subject of reserves and earnings, I became increasingly interested in the subject itself. What I have obtained so far is a "theory of accounting principles" rather than what I had originally sought. The main points of the theory are:

(a) There is a very close tie between classical "asset shares" and both Anderson's concept of a required IRR (= Internal Rate of Return) and GAAP (= Generally Accepted Accounting Principles). Different "accounting principles" lead in a surprisingly simple manner from asset shares to Anderson, GAAP, etc.

(b) It is necessary, however, to obtain a clear understanding of the nature of reserves as an accumulation process. The key to this understanding is adequate notation. In this paper, reserves V are a function V = S(B;i) where the "operator" S defines an accumulation process over book results B at interest i. The operator S is a generalization of "$\ddot{s}_{\overline{n}|i}$" and is similar to the "scan operator" in APL.

(c) We distinguish between reserve vectors and the reserve matrix. This is reminiscent of APL, but the concept of conforming arrays is not strictly applied to this theory. Rather, the reserve vector is thought of as an infinite row of zeros related to definite time periods. In the course of the accumulation process some zeros are replaced by other numbers.

(d) Reserve vectors are produced by the book results at the end of successive accounting periods.

(e) The reserve matrix is generated by accumulation of all cash flow items, arranged in a matrix such that a column shows the accumulation balance after the same kind of transaction, and a row covers one full accounting period.

(f) We need not distinguish between terminal reserves, mean reserves, etc. Reserves referring to different definitions of the accounting period emerge automatically from the accumulation process, and are represented by different columns of the reserve matrix.

(g) Contrary to common belief, statutory reserves need not be "ignored" when calculating GAAP reserves.

The results obtained so far are presented in five parts of the paper, written in different languages:

Part 1 derives the concept of accounting principles from a simple numerical example.

Part 2 contains the theory of reserve vectors.

Part 3 is an APL demonstration.

Part 4 introduces the reserve matrix, which simply combines reserve vectors column by column.

Part 5 reviews a concrete example.

The author would like to express sincere thanks to his friend, Brian Fortier, FSA, for a critical review of the manuscript and for many valuable suggestions.

PART 1. ACCOUNTING PRINCIPLES

Summary: Starting from IRR (internal rate of return) and asset share accumulations, the concept of accounting principles is developed. The terms "profit" versus "gain" are introduced.

Language: English supported by numerical illustrations.

Vocabulary:

B = Balance of cash transactions = Bookresult

i = Interest rate assumed

IRR = Internal rate of return

A = Asset share accumulation

G = Gain from operation before amortization reserve adjustment

V = Amortization reserve

π = Realized profit

γ = Adjusted gain from operations

1.1. *Profit and Gain*

Table 1 contains the amortization schedule for a $202,886 loan which is repayable in three decreasing installments. The effective rate of interest is 12%.

This example is adapted from the paper by Donald Sondergeld (TSA XXVI). Sondergeld does not consider a loan, but the cash flow of a four-year life insurance plan. Thus, the "loan" of $202,886 represents the first policy year book loss, and the "repayments" are book profits of the subsequent policy years. Obviously, the internal rate of return (IRR) of this business is 12%.

The IRR is one important tool to measure the profitability of a life insurance business; the concept was developed by James Anderson (TSA XI). Another such tool is asset share accumulation as demonstrated in *Table 2*. Asset share accumulation requires as additional assumption the *interest rate* i which can be interpreted in different ways, for instance:

-i may represent the anticipated return on invested assets. If the accumulated fund is negative, then assets are reduced and certain alternate investments are "lost" due to investment in the life insurance business.

-i may be a utility parameter which measures the combined effect of market interest and inflation. Then the discount function $u = u(t;i)$ represents the utility attached to the cash transaction "1" at a time t.

If actual experience matches theoretical expectation, the gain from operations would be equal to the increase of the asset share fund. Thus, we distinguish between:

Book Result $_tB$ which represents the balance of all cash transactions during the year t minus the increase of statutory reserves, if any. Each transaction may bear interest up to the end of year t. However, $_tB$ shall *not* include interest on previous book results $_kB$ $(k < t)$.

Asset share fund $_tA$ which is the accumulation of book results $_tB$ at the assumed interest rate i.

Gain from Operations $_tG = {}_tA - {}_{t-1}A$. Thus, gain $_tG$ equals book result $_tB$ plus interest on previous book results $_kB$ $(k < t)$.

Tables 1 and 2 are based on the same numerical example, but they have little in common. The most disturbing fact is the different sign of the fund: The "outstanding principal" in Table 1 has a positive sign, which makes sense since it is an asset. The asset share fund A is an asset too, yet A has a negative sign in the initial years. The explanation of this paradox lies in the different treatment of the first year result. In Table 1, the $202,886 is thought of as an amortizable asset; in Table 2, the $202,886 represents a realized loss.

We may rearrange Tables 1 and 2 so that the difference becomes more transparent. *Table 3* is equivalent to Table 1. In order to make the same numerical example applicable in both the loan amortization model as well as the life insurance model, some new terms are introduced:

Realized Profit $_t\pi$. If the effective rate of return is 12%, and the "market rate" or reinvestment rate is i = 5%, then we can assume that the interest differential of 7% constitutes a profit which is realized at each time the 12% return is received.

Amortization Reserve $_tV$ which is the accumulation of the difference $_tB - {}_t\pi$ at the assumed interest i. In the case of Table 3, $_tV$ equals the "outstanding principal" except for the sign. We may conceive of a reserve as a liability, and consider a negative reserve an asset.

Adjusted Gain from Operations $_t\gamma$, namely the difference between (a) the increase of the asset share fund A and (b) the increase of the amortization reserve V. In other words, ${}_t\gamma$ equals ${}_tG$ less the reserve increment ${}_tV - {}_{t-1}V$. Thus, the reported gain from operations becomes ${}_t\gamma$ if the amortization reserve V is set up.

Table 4 is equivalent to Table 2. In contrast to Table 3, realized profits π are assumed to equal book results B. This represents the special case of "Statutory Accounting" as required by the insurance laws. Amortization reserves V are zero in Table 4 and not equal to statutory reserves, because the latter are according to the assumptions made already included in the book results B. Thus, amortization reserves V as defined in this chapter are actually differences from statutory reserves; we shall demonstrate later that this does not constitute a significant restriction.

When comparing Tables 3 and 4 we notice:

a) Realized profits ${}_t\pi$ are different, but they accumulate to the same amount, \$29,384, if reinvested at the assumed interest rate. As we can see from Table 2, \$29,384 is the amount which the asset share fund reaches at the end of the period under consideration.

b) The totals of ${}_t\gamma$, the gains from operations, amount also to \$29,384. Thus, the different accounting principles represented by Tables 3 and 4, respectively, merely redistribute the total gain from operations.

c) Amortization reserves ${}_tV$ go to zero at the end of the period under consideration.

We shall prove later that a) and b) are merely consequences of c). Thus, we may define an *Accounting Principle* as an assumption on the emergence of profits ${}_t\pi$, subject to the condition that the profits ${}_t\pi$ accumulate to the final pattern of the asset share fund, ${}_nA$, if they are reinvested at the assumed interest rate i.

Tables 3 and 4 illustrate two different accounting principles which may be called *IRR-Accounting* and *Statutory Accounting*, respectively.

As loans can be amortized under many different methods, there is a variety of accounting principles to report life insurance earnings. We chall consider two more examples.

In *Table 5*, the realization of profits is deferred to the end of the insurance period. As comparison with Table 2 reveals, the amortization reserves ${}_tV$ are in this case identical with the asset share fund ${}_tA$, except in the last year. Therefore, this (hypothetical) accounting principle may be denoted as *"Asset-share Accounting"*. The adjusted gain ${}_t\gamma$ is zero, except in the last policy year.

Finally, *Table 6* illustrates *GAAP Accounting*, that is, Generally Accepted Accounting Principles. Under this method, the profits ${}_t\pi$ are assumed to emerge as a uniform percentage of premium income ${}_tP$, except that certain "non-amortizable expenses" may be treated separately.

The most common term used in connection with GAAP is "adjusted earnings". However, the definition varies in different papers: it may refer to ${}_t\pi$ or to ${}_t\gamma$.

For instance, Pharr (1971) distinguishes between "adjusted earnings" π, and "adjusted earnings plus interest" γ. Paquin (1973) defines π as "adjusted earnings". Sondergeld (1975) uses "GAAP profits" for π and "GAAP earnings" for γ. Milgrom (1975) distinguishes "gain" π from "earnings" γ.

REFERENCES

Anderson, J.C.H. (1959). Gross premium calculation and profit measurement for non-participating insurance, TSA XI.

Pharr, Joe B. (1971). The natural reserve concept and life insurance earnings, TSA XXIII.

Paquin, Claude Y. (1974). The development of mean reserve factors and methods of amortizing acquisition expenses in adjusting life insurance company earnings, TSA XXV.

Sondergeld, Donald R. (1975). Earnings and the internal rate of return measurement of profit, TSA XXVI.
Milgrom, Paul R. (1975). On understanding the effects of GAAP reserve assumptions, TSA XXVII.

TABLE 1. Amortization Schedule at 12% Interest

End of Year t	Loan (1)	Repay-ments (2)	12% on Principal (3)	Amorti-zation (2)-(3)	Outstanding Principal (5)
1	202,886		0	0	202,886
2		100,000	24,346	75,654	127,232
3		80,000	15,268	64,732	62,500
4		70,000	7,500	62,500	0
Totals:	202,886	250,000	47,114	202,886	

TABLE 2. Asset-share Accumulation Interest i = 5%

End of Year t	Book Result ${}_tB$	Interest on Fund ${}_tI{-}i\,{}_{t-1}A$	Increase of Fund ${}_tB+{}_tI$	Asset share Fund ${}_tA$	Gain from Operations ${}_tG={}_tA-{}_{t-1}A$
1	-202,886	0	-202,886	-202,886	-202,886
2	100,000	-10,144	89,856	-113,030	89,856
3	80,000	- 5,652	74,348	- 38,682	74,348
4	70,000	- 1,934	68,066	29,384	68,066
Totals:	47,114	-17,730	29,384		29,384

TABLE 3. *Amortization Schedule for IRR-Accounting*
Internal rate of return: 12%
Interest i = 5%

End of Year t	Book-Result ${}_tB$	Realized Profit $-7\%\ {}_{t-1}V = {}_t\pi$	Interest on Reserve $5\%\ {}_{t-1}V = {}_tI$	Reserve Increase ${}_tB - {}_t\pi + {}_tI$	Amortization Reserve ${}_tV$	Gain from Operations ${}_t\gamma = {}_tG - {}_tV + {}_{t-1}V$
1	-202,886	0	0	-202,886	-202,886	0
2	100,000	14,202	-10,144	75,654	-127,232	14,202
3	80,000	8,906	6,362	64,732	- 62,500	9,616
4	70,000	4,375	- 3,125	62,500	0	5,566
Totals:	47,114	27,483	-19,631	0		29,384
Accum.*	29,384	29,384	-20,989	-20,989		31,321

*Accumulation with 5% interest

TABLE 4. Amortization Schedule for Statutory Accounting
Interest $i = 5\%$

End of Year t	Book-Result ${}_tB$	Realized Profit ${}_t\pi$	Interest on Reserve $i \cdot {}_{t-1}V = {}_tI$	Reserve Increase ${}_tB - {}_t\pi + {}_tI$	Amortization Reserve ${}_tV$	Gain from Operations $\gamma_t = {}_tG - {}_tV + {}_{t-1}V$
1	-202,886	-202,886	0	0	0	-202,886
2	100,000	100,000	0	0	0	89,856
3	80,000	80,000	0	0	0	74,348
4	70,000	70,000	0	0	0	68,066
Totals:	47,114	47,114	0	0	0	29,384
Accum.*	29,384	29,384	0	0	0	10,332

*Accumulation with 5% interest

TABLE 5. Amortization Schedule for "Asset-share Accounting"
Interest $i = 5\%$

End of Year t	Book-Result $_tB$	Realized Profit $_t\pi$	Interest on Reserve $i \cdot {}_{t-1}V = {}_tI$	Reserve Increase $_tB - {}_t\pi + {}_tI$	Amortization Reserve $_tV$	Gain from Operations $_t\gamma = {}_tG - {}_tV + {}_{t-1}V$
1	-202,886	0	0	-202,886	-202,886	0
2	100,000	0	-10,144	89,856	-113,030	0
3	80,000	0	- 5,652	74,348	- 38,682	0
4	70,000	29,384	- 1,934	38,682	0	29,384
Totals:	47,114	29,384	-17,730	0		29,384
Accum.*	29,384	29,384	-19,052	-19,052		29,384

*Accumulation with 5% interest

TABLE 6. *Amortization Schedule for GAAP Accounting*
Interest i = 5%
Profit realized in relation 100:75:65:60

End of Year t	Book-Result ${}_tB$	Realized Profit ${}_t\pi$	Interest on Reserve $i \cdot {}_{t-1}V = {}_tI$	Reserve Increase ${}_tB - {}_t\pi + {}_tI$	Amortization Reserve ${}_tV$	Gain from Operations ${}_t\gamma = {}_tG - {}_tV + {}_{t-1}V$
1	-202,886	8,994	0	-211,880	-211,880	8,994
2	100,000	6,746	-10,594	82,660	-129,220	7,196
3	80,000	5,846	- 6,461	67,693	- 61,527	6,655
4	70,000	5,397	- 3,076	61,527	0	6,539
Totals:	47,114	26,983	-20,131	0		29,384
Accum.*	29,384	29,384	-21,540	-21,540		31,872

*Accumulation with 5% interest

PART 2. RESERVE VECTORS

Summary: Rigorous treatment of accounting principles.

Language: Algebra

Dictionary:

B	Bookresult-Vector
i	Interest-Vector
$A = S(B,i)$	Asset share-Vector
$G = \nabla A$	Vector of gains from operation
$V = S(B-\pi;i)$	Vector of amortization reserves
π	Vector of realized profits, subject to the condition
${}_nS(\pi;i) = {}_nA$	
$\gamma = \nabla A - \nabla V$	Vector of adjusted gains from operation

2.1. *Model*

Let B be a vector of *book results*, which emerge at the end of n insurance periods (say years):

$$B = {}_1B,\ {}_2B,\ \ldots,\ {}_{n-1}B,\ {}_nB.$$

Let R be a vector of *reserves* at the end of each insurance period:

$$R = {}_1R,\ {}_2R,\ \ldots,\ {}_{n-1}R,\ {}_nR.$$

We shall later consider as special cases of R:

A the vector of asset-share funds,

V the vector of amortization reserves.

The *interest* prevailing in the insurance periods shall be represented by the vector

$$i = {}_1i,\ {}_2i,\ \ldots,\ {}_{n-1}i,\ {}_ni.$$

Finally, we define the *discount function* $u(t)$:

$$\begin{cases} u(1) = (1+{}_1i)^{-1} \\ u(t) = u(t-1) \cdot (1+{}_ti)^{-1} \text{ for } t > 1 \end{cases}$$

In case of constant interest i we thus have:

$$u(t) = (1+i)^{-t}.$$

Note: $u(t)$ is an utility function putting weight on the timing of transactions.

2.2. *Manipulation of Vectors*

The main advantage of using vectors is convenience; that is, we are spared unneeded indices.

Let us agree on the following definitions which may be applied to any vectors, say R and R'.

Equal Sign: $R' = R$ shall mean that

$${}_tR' = {}_tR \text{ for all } t \ (1 \leq t \leq n).$$

Backshift Operator:

If $R = {}_1R,\ {}_2R,\ \ldots,\ {}_{n-1}R,\ {}_nR,$

then ${}_*R = 0,\ {}_1R,\ \ldots,\ {}_{n-2}R,\ {}_{n-1}R.$

Backward-Difference Operator: $\nabla R = R - {}_*R.$

That is, $\nabla R = \nabla_1 R,\ \nabla_2 R,\ \ldots,\ \nabla_{n-1} R,\ \nabla_n R,$

where $\begin{cases} \nabla_1 R = {}_1R \\ \nabla_t R = {}_tR - {}_{t-1}R \text{ for } t > 1. \end{cases}$

2.3. *Reserve Accumulation*

We shall generally apply the retrospective method. Thus, the reserve accumulation can be symbolized by the

Accumulation Operator $S(B;i)$:

$R = S(B;i)$ shall mean that

${}_tR = {}_tS(B;i)$ for all $t \geq 1$

where

$${}_tS(B;i) = \frac{1}{u(t)} \sum_{k=1}^{t} u(k) \cdot {}_kB.$$

Examples. If ${}_ti = \text{const} = i$, then

$${}_tS(B;i) = (1+i)^t \sum_{k=1}^{t} {}_kB \cdot (1+i)^{-k}$$

If ${}_ti = \text{const} = i$ and ${}_tB = \text{const} = 1$, then

$${}_tS(1;i) = (1+i)^t \sum_{k=1}^{t} (1+i)^{-k} = \ddot{s}_{\overline{t|}i}.$$

If ${}_ti = \text{const} = 0$, then

$${}_tS(B;0) = \sum_{k=1}^{t} {}_kB.$$

Note. The "initial reserve" ${}_0R$ is not defined, since accumulation starts with ${}_1B$.

However, we may put ${}_0R = 0$ as tacitly assumed when defining ${}_*R$ and ∇R. See item (c) of the introduction to this paper.

2.4. *The Accumulation Operator $R = S(B;i)$ as a Function of B and i*

Variation of B:

If $R = S(B;i)$

and $R' = S(B';i)$

then $R-R' = S(B-B';i)$. (1)

Proof. $S(B;i)$ is additive with respect to B.

"Solving" for B:

If $R = S(B;i)$ then $B = \nabla R - i \cdot {}_{*}R.$ (2)

Proof. By definition of $R = S(B;i)$

$$u(t) \cdot {}_{t}R = u(1) \cdot {}_{1}B + \ldots + u(t-1) \cdot {}_{t-1}B + u(t) \cdot {}_{t}B$$

$$u(t-1) \cdot {}_{t-1}R = u(1) \cdot {}_{t}B + \ldots + u(t-1) \cdot {}_{t-1}B.$$

Thus,

$$u(t) \cdot {}_{t}B = u(t) \cdot {}_{t}R - u(t-1) \cdot {}_{t-1}R.$$

Divide by $u(t)$ and observe that $1 + {}_{t}i = u(t-1)/u(t)$:

$${}_{t}B = {}_{t}R - (1+{}_{t}i) \cdot {}_{t-1}R = \nabla_{t}R - {}_{t}i \cdot {}_{t-1}R.$$

Equivalent arguments of S:

If $R = S(B;i) = S(B';i')$ then $B' = B + (i-i') \cdot {}_{*}R.$ (3)

Proof. From (2) it follows that $B = \nabla R - i \cdot {}_{*}R$ and $B' = \nabla R - i' \cdot {}_{*}R$. Thus, $B' + i \cdot {}_{*}R = B + i' \cdot {}_{*}R$.

Identity:

If $R = S(B;i)$ then $R = S(B+i \cdot {}_{*}R;0) = S(\nabla R;0).$ (4)

Proof. Put $i' = 0$ in (3) and observe (2).

Variation of interest i:

If $R = S(B;y)$ then $R = S(B+[y-i] \cdot {}_{*}R;i).$ (5)

Proof. Replace i by y and i' by i, apply (3).

2.5. *Evaluation of a Life Insurance Business*

Let

$$B = {}_{1}B,\ {}_{2}B,\ \ldots,\ {}_{n}B$$

represent the book profits which emerge according to actuarial projections from a collection of life insurance policies, issued at the same time under one plan. B shall include Statutory reserve increases, but not interest on previous book profits.

Assume that generally ${}_1B < 0$ because of acquisition costs, but that generally ${}_tB > 0$. Then the equation

$$ {}_nS(B;y) = {}_1B.(1+y)^{n-1} + \dots + {}_nB.(1+y)^0 = 0 $$

has a solution y = IRR, which is the *Internal Rate of Return*. The IRR is one important measure of profitability; the concept was developed by J.C.H. Anderson (TSA XI, 1959).

Another tool to measure profitability is the *Asset-share Accumulation*

$$ A = S(B;i). \tag{6} $$

In order to calculate asset shares, we have to make an assumption on the *interest i*, which complements all the other assumptions which were necessary to determine B.

The expected *Gains from Operations* arising from the block of business under consideration are

$$ G = \nabla A = B + i\cdot {}_*A. $$

The first of these two equations defines gain G as the increase ∇A of the asset share fund. The latter equation is a consequence of (2) and tells that the total expected gain from operations consists of

Book profits B, plus

Interest income from the fund, i.e., from book profits of previous years.

The profitability of the business may be measured by

$$ {}_nA = {}_nS(G;0) = {}_nS(B;i), $$

namely the sum of all gains, which is equal to the accumulation of all book profits B, reinvested at interest i.

2.6. *Reporting of Life Insurance Earnings*

In essence, an *Accounting Principle* is a set of rules which determine the

adjusted gains γ from operations, thereby requiring, as a balancing item,

amortization reserves V which are to be set up in addition to Statutory reserves (but generally have negative sign). These amortization reserves can be thought of as produced by recognition of

emerging profits π.

These three entities are related to each other, and to the asset share fund $A = S(B;i)$, as follows:

$$A - V = S(\pi;i) = S(\gamma;0), \text{ where } {}_nV = 0. \tag{7}$$

From the condition ${}_nV = 0$ follows:

$${}_nS(\pi;i) = {}_nS(\gamma;0) = {}_nA. \tag{8}$$

From (1), (2) and (7):

$$\gamma = \nabla A - \nabla V. \tag{9}$$

That is, gain equals increase in assets less reserve increase.

From (1) and (7):

$$V = A - S(\pi;i) = S(B;i) - S(\pi;i) = S(B-\pi;i). \tag{10}$$

Thus, amortization reserves V can be interpreted as accumulation of book results B less emerging profits π.

Finally, from (3) and (7):

$$\gamma \equiv \pi + i\cdot{}_*R, \text{ where } R = S(\pi;i) = A - V. \tag{11}$$

That is, gains ${}_t\gamma$ are equal to profits ${}_t\pi$ plus interest on previous profits ${}_k\pi$ $(k < t)$.

2.7. *Special Accounting Principles*

The following examples demonstrate that an accounting principle may be expressed in any of the entities π, γ or V. One could say that the following examples refer to those accounting principles which satisfy the conditions ${}_nV = 0$ or (8) in the most straightforward manner.

Statutory Accounting: $\pi = B$. Thus,

$$V = S(B-\pi;i) = S(0;i) = 0$$

and

$$\gamma = \nabla A - \nabla V = \nabla A = G.$$

Asset-share Accounting: $\begin{cases} {}_t\gamma = 0 & \text{if } t < n, \\ {}_n\gamma = {}_nA. \end{cases}$

$$\begin{cases} {}_tV = {}_tA - {}_tS(\gamma;0) = {}_tA - 0 = {}_tA & \text{if } t < n; \\ {}_nV = {}_nA - {}_n\gamma = {}_nA - {}_nA = 0 \end{cases}$$

From (11):

$$\begin{cases} {}_t\pi = {}_t\gamma - i\cdot({}_{t-1}A-{}_{t-1}V) = 0 - i\cdot 0 = 0 & \text{if } t < n; \\ {}_n\pi = {}_n\gamma - i\cdot({}_{n-1}A-{}_{n-1}V) = {}_nA - i\cdot 0 = {}_nA. \end{cases}$$

IRR - Accounting: $V = S(B;y)$ where ${}_nV = {}_n(SB;y) = 0$.

We recall (4) and (5), namely

$$V = S(B;y) = S(B+y\cdot{}_*V;0) = S(B+[y-i]\cdot{}_*V;i).$$

By comparison with (7) we find

$$\pi = B - (B+[y-i]\cdot{}_*V) = (i-y)\cdot{}_*V;$$

$$\gamma = \pi + i\cdot{}_*(A-V) = i\cdot{}_*A-y\cdot{}_*V.$$

Thus, IRR profits are equal to the excess interest on IRR amortization reserves, taken at opposite sign where generally $V < 0$.

2.8. *Generally Accepted Accounting Principles (GAAP)*

Let the book result B be the balance

$$B = P - BE - AE - NE$$

with

P = Gross premium income

BE = all benefits except expenses

AE = amortizable expenses

NE = non-amortizable expenses.

Under GAAP the total amortization reserve V is the sum of benefit reserve V^B and expense reserve V^E:

$$V = V^B + V^E \text{ where } {}_nV^B = {}_nV^E = 0.$$

From (7) it follows that

$$A - V = A - V^B - V^E = S(\pi;i) = S(\gamma;0). \tag{12}$$

The *benefit reserve* is

$$V^B = S(\beta P - BE;i), \tag{13}$$

where the constant factor β must satisfy:

$$\beta \cdot {}_nS(P;i) = {}_nS(BE;i).$$

As to the *expense reserve*, we have the alternative between

a) direct assumptions on V^E, so that $V = V^B + V^E$,

b) assumptions on π which determine V, so that $V^E = V - V^B$.

Alternative a. Direct assumptions on V^E are applied under various permissible *Expense-Runoff-Methods* (see Paquin), subject only to

$${}_nV^E = 0.$$

Alternative b. It shall be assumed that profits will emerge as a constant percentage of gross premium income P, except that non-amortizable expenses NE are charged immediately against income. Hence,

$$\pi = \lambda P - NE.$$

From (10) it follows that

$$V = S(B-\pi;i) = S(B+NE-\lambda P;i), \quad (14)$$

where the constant factor λ is determined by the condition ${}_nV = 0$:

$$\lambda \cdot {}_nS(P;i) = {}_nS(B+NE;i).$$

To reconcile, we have from (13):

$$V^B = S(\beta P - BE;i),$$

and from (14) less (13):

$$V^E = S(\varepsilon P - AE;i), \text{ where } \varepsilon = 1 - \lambda - \beta.$$

2.9. *Extension of Notation*

In order to present an analysis of accounting principles, we have made extensive use of some mathematical tools, namely

the vector concept

the "backshift operator"

the backward difference ∇

the "accumulation operator" S

The two operators put in quotes are defined for the purposes of this paper. Standard finite difference texts use the symbols E^{-1} or B for the backshift operation.

The first three tools listed serve mainly the purpose of convenience.

The accumulation operator S plays a key role, namely as a means of communication and as a tool for logical conclusions. The operator S (which could be viewed as an extension of the $\sum$ symbol) is implemented in APL as the "Scan-Operator".

It is tempting to introduce also a *partial inverse of S*, say $\underline{S}$:

$$B = \underline{S}(R;i) \text{ if } R = S(B;i).$$

Hence,

$$\underline{S}(R;i) = \nabla R - i \cdot {}_*R.$$

An alternative notation would be either S^{-1} or ∇:

$$\nabla(R;i) = \nabla R - i \cdot_* R \text{ with } \nabla(R;0) = \nabla R.$$

It was felt that the introduction of $\underline{S}$ would have overloaded this presentation. However, for computational purposes $\underline{S}$ seems indispensable; therefore $\underline{S}$ was utilized in the APL version.

PART 3. APL DEMONSTRATION

Summary: Computational varification of the examples presented in Part 1.

Language: APL

Dictionary:

B	B
i	I
$A = S(B;i)$	A ← I S B
$G = \nabla A$	G ← 0 $\underline{S}$ A
$V = S(B-\pi;i)$	V ← I S B-PI
$\pi = \underline{S}(A-V;i)$	PI ← I $\underline{S}$ A-V
$\gamma = \underline{S}(A-V;0)$	G → 0 $\underline{S}$ A-V
${}_tV$	V[T] = element t of vector V.

3.1. *Computer Terminal Session*

The computer printout which follows first shows the programs for the two functions S and $\underline{S}$.

Function Name	Syntax of Function	Result Equals
S	R ← I S B	$R = S(B;I)$
$\underline{S}$	B ← I $\underline{S}$ R	$B = \underline{S}(R;I) = R - I \cdot {*}R$

Next, Table 2 is verified in the following steps:

Vector B is created, given an internal rate of return of 12%.

Vector A is calculated, given B and an interest rate of 5%.

Vector G (containing Gains from Operations) is generated.

Vector B is recalculated, given A and an interest rate of 5%.

Vector A is recalculated in two different ways; the results agree because of the identity:

$$A = S(B;i) = S(G;0).$$

Finally, Tables 3 through 6 are reproduced in the following order:

Amortization reserves V, given the definition of the respective accounting principles.

Gains γ	$\equiv$	$0\ \underline{S}$ A-V
Profits π	$\equiv$	$.05\ \underline{S}$ A-V
Accumulation $S(\gamma;0)$	$\equiv$	0 S GAMMA
Accumulation $S(\pi;i)$	$\equiv$	.05 S PI.

Due to equation (7), the last two steps yield the same vector.

```
        ⍝ *****   PROGRAMS   *****

        ∇S[⎕]∇
      ∇ V←I S B;U
[1]     V←(+\B×U)÷U←×\(⍴B)⍴÷1+I
[2]   ⍝ "V←I S B" ←→ V=S(B;I)
      ∇
        ∇S̲[⎕]∇
      ∇ B←I S̲ V
[1]     B←V-(1+I)×(⍴V)↑0,[1] V
[2]   ⍝ SOLVES  V=S(B;I)  FOR B
      ∇

    ⍝ *****  ASSETSHARE FUND A  *****  (TABLE 2)

    10 3⍕  B  ← (-B+.÷1.12* 1 2 3),B← 100 80 70
¯202.886   100.000     80.000     70.000

    10 3⍕  A  ←  .05  S  B
¯202.886  ¯113.030    ¯38.682     29.384

    10 3⍕  G  ←    0  S̲  A
¯202.886    89.856     74.348     68.066
    10 3⍕  B  ←  .05  S̲  A
¯202.886   100.000     80.000     70.000

    10 3⍕    0  S  G
¯202.886  ¯113.030    ¯38.682     29.384
    10 3⍕  .05  S  B
¯202.886  ¯113.030    ¯38.682     29.384
```

```
⍝ *****  IRR-ACCOUNTING  *****  (TABLE 3)

   10 3⍕  V  ←  .12  S  B
¯202.886  ¯127.232   ¯62.500       .000

   10 3⍕ GAMMA ←  0  S  A-V
 ¯.000     14.202     9.616      5.566
   10 3⍕ PI  ←  .05  S  A-V
 ¯.000     14.202     8.906      4.375

   10 3⍕     0  S  GAMMA
 ¯.000     14.202    23.818     29.384
   10 3⍕  .05  S  PI
 ¯.000     14.202    23.818     29.384

⍝ *****  STATUTORY ACCOUNTING  *****  (TABLE 4)

   10 3⍕  V  ←  .05  S  (B - PI←B)
  .000       .000       .000       .000

   10 3⍕ GAMMA ←  0  S  A-V
¯202.886    89.856    74.348     68.066
   10 3⍕ PI  ←  .05  S  A-V
¯202.886   100.000    80.000     70.000

   10 3⍕     0  S  GAMMA
¯202.886  ¯113.030   ¯38.682     29.384
   10 3⍕  .05  S  PI
¯202.886  ¯113.030   ¯38.682     29.384

⍝ *****  ASSETSHARE ACCOUNTING  *****  (TABLE 5)

   10 3⍕  V ←  .05  S  (B - PI← 0 0 0 ,A[4] )
¯202.886  ¯113.030   ¯38.682      ¯.000

   10 3⍕ GAMMA ← 0  S  A-V
 ¯.000       .000       .000     29.384
   10 3⍕ PI  ← .05  S  A-V
 ¯.00        .000       .000     29.384

   10 3⍕     0  S  GAMMA
 ¯.000       .000       .000     29.384
   10 3⍕  .05  S  PI
 ¯.000      ¯.000       .000     29.384

⍝ *****  GAAP-ACCOUNTING  *****  (TABLE 6)

   PREM← 100 75 65 60
   10 3⍕ PI ← PREM×A[4]÷PREM+.×1.05* 3 2 1 0
  8.994      6.746      5.846      5.397

   10 3⍕  V  ←  .05  S  (B-PI)
¯211.880  ¯129.220   ¯61.527      ¯.000

   10 3⍕ GAMMA ←  0  S  A-V
  8.994      7.195      6.656      6.539
   10 3⍕ PI  ←  .05  S  A-V
  8.994      6.746      5.846      5.397

   10 3⍕     0  S  GAMMA
  8.994     16.190     22.845     29.384
   10 3⍕  .05  S  PI
  8.994     16.190     22.845     29.384
```

PART 4. THE RESERVE MATRIX

Summary: The concept of a reserve matrix is developed. Each column of the reserve matrix is a reserve vector in the definition of Part 2. Thus, the conclusions drawn in Part 2 may simply be applied to the rows of the reserve matrix.

Language: Algebra.

Dictionary:

C	Matrix of cashflow transactions
C'	Matrix of cashflow transactions, adjusted for assumed realized profits π_p.
i	Interest vector
A = R(C)	Asset share matrix
B	Matrix of annual book results, where A = S(B;i)
V = R(C')	Matrix of amortization reserves
π_p	Vector of assumed realized profits, where ${}_nS(\pi p;i) = {}_nA_p$
π	Matrix of realized profits, where $S(\pi;i) = A - V$
$\gamma = \nabla A - \nabla V$	Matrix of adjusted gains from operations.

4.1. *Concept*

So far we have treated reserves as vectors R with the components

$${}_tR = {}_tS(B;i).$$

Thus, the reserve vector R is a function of the vector B of book results.

Actually, the vector B depends on a matrix C representing cashflow transactions such that

$${}_tB = \sum_{h-1}^{z} {}_tC_h \cdot (1+{}_ti)^{\tau(h)}.$$

This equation says that the book result of the year t, ${}_tB$, is the balance of z single transactions ${}_tC_h$ (h = 1...z), carried with interest to the year end.

Obviously, we can arrange the matrix $\{{}_tC_h\}$ in different ways, for example by policy year or by calendar year. The annual book results, ${}_tB$, depend on the particular arrangement of the C matrix. Thus, we may write

$$B_z = {}_1B_z, \; {}_2B_z, \; \ldots, \; {}_nB_z$$

for the particular vector of book results which arise if the accounting periods are such that they end with transaction z.

Accordingly, we have different reserve *vectors*

$$R_z = S(B_z; i)$$

which we may combine, column by column, into a *matrix* $\{{}_tR_z\}$.

Thus, different columns of $\{{}_tR_z\}$ contain reserves for different accounting periods, say

terminal reserves (at the end of the policy years)

mean reserves (at the end of calendar years)

reserves before or after such transactions as setting up statutory reserves.

4.2. *The Cashflow Matrix C*

Consider a 3-year term plan involving the following cashflow transactions:

Policy year :	1	2	3
Premiums :	P1	P2	P3
Expenses :	E1	E2	E3
Death benefits:	D1	D2	D3.

These transactions occur in a certain *order of events*, say

P1 E1 D1, P2 E2 D2, P3 E3 D3.

We are mainly interested in book results of given accounting periods. Therefore, we may find it more convenient to arrange the transactions in matrix form. We shall call such arrangement a *cash-flow matrix*. Shown below are two examples, depending on which accounting period is chosen.

Policy-year basis:

P1	E1	D1
P2	E2	D2
P3	E3	D3

Calendar-year basis:

0	P1	C1
D1	P2	C2
D2	P3	C3
D3	0	0

These two matrices have in common:

a) They are generated from the same order of events. This means: if read row by row, each cash-flow matrix yields the same vector, where zeros at the beginning and the end of the vector are ignored.

b) Each row combines the transactions of the same accounting period. Thus, book results ${}_tB$ are the balance of all items contained in row t.

c) Each column contains items referring to the same kind of transaction.

Suppose a certain order of events (which remains invariant) is given. From this order we can derive a set of cash-flow matrices C, depending on which accounting period is chosen. Each matrix C determines one vector B of book results.

Let ${}_tC_h$ denote an element of C, where

t the row number represents accounting periods ($1 \leq t \leq n$),
h the column number represents kinds of transactions ($1 \leq h \leq z$).

In order to avoid index confusion, we shall pick a particular matrix C which remains invariant too.

This does not prevent us from considering the whole set of book results which could be derived from the order of events. Indeed, we may define ${}_tB_h$ as the book result of the (fictitious) accounting period which terminates after event (t,h). Thus, B_z

defines the vector of book results which corresponds to the particular matrix C we have chosen.

The term "cash-flow transactions" shall also cover the setting up and the release of Statutory reserves.

Finally, the sign of ${}_tC_h$ shall serve to distinguish between income (+) and expenditure (-).

4.3. *The Discount Function u(t,h)*

Let us agree on a time scale by imposing:

a) the length of one accounting period is the time unit.

b) the event (t,z), where z represents the last column of C, occurs at time t.

c) the time difference between events (t,h) and (t,z) is $\tau(h)$. Obviously, $0 \le \tau \le 1$.

We may now extend the definition of the *discount function* u(t,h) as follows:

$$u(t,h) = (1+i)^{\tau(h)-t} \quad \text{if} \quad i = \text{constant}.$$

More generally, if ${}_ti$ denotes the rate of interest along one row of C, we impose:

$$\begin{cases} u(1,z) = (1+{}_1i)^{-1} \\ u(t,z) = u(t-1,z) \cdot (1+{}_ti)^{-1} \text{ if } t > 1 \\ u(t,h) = u(t,z) \cdot (1+{}_ti)^{\tau(h)}. \end{cases}$$

Thus, the *book result* ${}_tB_z$ of a full accounting period which ends with event (t,z) is

$${}_tB_z = \sum_{h=1}^{z} {}_tC_h \cdot (1+{}_ti)^{\tau(h)} = \sum_{h=1}^{z} {}_tC_h \cdot u(t,h)/u(t,z). \quad (15)$$

4.4. *Reserve Accumulation*

In accord with the reserve definition of Part 2, Chapter 2.3, we have

$$ {}_tR_z = {}_tS(B_z;i) = \frac{1}{u(t,z)} \sum_{k=1}^{t} u(k,z) \cdot {}_kB_z. \tag{16} $$

Inserting the book values ${}_kB_z$ from (15) we find

$$ {}_tR_z = \frac{1}{u(t,z)} \sum_{k=1}^{t} u(k,z) \cdot \sum_{h=1}^{z} {}_kC_h \; u(k,h)/u(k,z) $$

$$ = \frac{1}{u(t,z)} \sum_{k,h} u(k,h) \cdot {}_kC_h, \tag{17} $$

where the sum in the second line of (17) is extended row after row of the C-matrix up to and including transaction ${}_tC_z$.

We learn from (17) that we need not determine first book results, then reserves. Instead, we can generate reserves directly from the cashflow items ${}_kC_h$ by means of a linear accumulation process on the order of events.

There is no reason to terminate the accumulation after selected events (1,z), (2,z), ... only. Obviously, the process can be interrupted after any step. Thus, we may define in much greater generality:

$$ {}_tR_h = \frac{1}{u(t,z)} \sum_{k,m} u(k,m) \cdot {}_kC_m, \tag{18} $$

where the sum is extended on the order of events (row by row) from transaction ${}_1C_1$ through transaction ${}_tC_h$.

Equations (18) define a *reserve-matrix* R with elements ${}_tR_h$. R is a function of C,

$$ R = R(C), $$

in the following sense: First, the elements of R are generated by accumulation of the elements of C; second, matrices R and C have identical structure, meaning that corresponding elements ${}_tR_h$ and ${}_tC_h$ have the same coordinates.

Obviously, each element ${}_{t}R_{h}$ of the reserve-matrix R defines the reserve at the end of the accounting period which terminates after event (t,h).

Thus, each column of the reserve matrix represents a vector of reserves, where each such vector refers to another definition of the accounting period. For instance, we will find in different columns of R

Initial reserves,

Terminal reserves,

Mean reserves.

4.5. *Manipulations Along Columns of the Reserve Matrix R*

The reserve matrix R is a combination of several reserve vectors, assembled column by column.

Thus, the theory of reserve vectors presented in Part 2 applies to any column of the reserve matrix R.

We shall adopt the notation of Part 2 with the understanding that the vector operators (introduced in Chapters 2.2 and 2.3) are to be executed simultaneously along all columns of the matrix. That implies in particular:

Accumulation - Operator S:

$R = S(B;i)$ shall mean that ${}_{t}R_{h} = {}_{t}S(B_{h};i)$ for all $t > 1$, where

$$ {}_{t}S(B_{h};i) = \frac{1}{u(t,h)} \sum_{k=1}^{t} u(k,h) \cdot {}_{k}B_{h}. $$

"Solving" for B: $R = S(B;i)$ implies that

$$ {}_{t}B_{h} = {}_{t}R_{h} - (1+{}_{t}i) \cdot {}_{t-1}R_{h}. \tag{19} $$

This follows from equation (2).

4.6. *Manipulations Along the Rows of the Reserve Matrix*

How are reserves ${}_tR_h$ and ${}_tR_{h-1}$ related?
We find from (18):

$$ {}_tR_{h-1} \cdot u(t,h-1) = \ldots + u(t,h-1) \cdot {}_tC_{h-1} $$

$$ {}_tR_h \cdot u(t,h) = \ldots + u(t,h-1) \cdot {}_tC_{h-1} + u(t,h) \cdot {}_tC_h $$

Thus,

$$ {}_tR_{h-1} \cdot u(t,h-1) = ({}_tR_h - {}_tC_h) \cdot u(t,h), $$

or

$$ \begin{aligned} {}_tR_h &= {}_tC_h + {}_tR_{h-1} \cdot u(t,h-1)/u(t,h) \\ &= {}_tC_h + {}_tR_{h-1} \cdot (1+{}_ti)^{\delta}. \end{aligned} \tag{20} $$

That is, reserves ${}_tR_h$ equal transactions ${}_tC_h$ plus the preceding reserve ${}_tR_{h-1}$ times an interest factor, where

$$ \delta = \tau(h-1) - \tau(h) $$

represents the time difference between events (t,h-1) and (t,h).

4.7. *Asset Shares A and Amortization Reserves V*

We may write

$$ A = R(C), $$
$$ V = R(C'). $$

This shall mean that the asset share matrix A and the matrix V of amortization reserves are both defined by equations (18), where C represents the cashflow matrix. C' equals C except for emerging profits.

Assuming that profits emerge at transactions p, we have

$$ \begin{cases} {}_tC'_h = {}_tC_h & \text{if } h \neq p, \\ {}_tC'_p = {}_tC_p - {}_t\pi_p. \end{cases} \tag{21} $$

Here, π_p represents the vector of emerging profits, defined in accordance with condition (8) such that

$${}_nS(\pi_p;i) = {}_nA_p.$$

We must distinguish between

a) the input vector π_p which defines a matrix V of amortization reserves, and

b) the output matrix π obtained by solving $A - V = S(\pi;i)$. Different columns π_h represent emerging profits for different accounting periods.

The concept of profits emerging at specific times applies also to Statutory accounting. Assuming $p = z$ we require $\pi_z = B_z$, where B_z is determined from $A_z = S(B_z;i)$. From definition $V = S(B-\pi;i)$ follows $V_z = 0$, in agreement with Part 2, Chapter 2.7. This does *not* imply $V = 0$ for the whole matrix. Rather, the intermediate reserves ${}_tV_h$ develop from formulae (20).

4.8. *Profits* π *and Gains* γ

Suppose a matrix V of amortization reserves is given, subject to the assumption that profits π_p emerge at transaction p. Then, emerging profits π and adjusted gains γ for all possible accounting periods are defined by formulae (7) from Part 2:

$$A - V = S(\pi;i) = S(\gamma;0).$$

To solve for π and for γ, we apply (19) and find in case of constant interest rate i:

$${}_t\pi_h = {}_tR_h - (1+i)\,{}_{t-1}R_h \quad \text{where } R = A - V$$

$${}_t\pi_{h-1} = {}_tR_{h-1} - (1+i)\,{}_{t-1}R_{h-1}$$

and

$${}_t\gamma_h = {}_tR_h - {}_{t-1}R_h \quad \text{where } R = A - V$$

$${}_t\gamma_{h-1} = {}_tR_{h-1} - {}_{t-1}R_{h-1}$$

On the other hand, we find from (20) and (21):

$$ {}_tR_h = {}_tR_{h-1} \cdot (1+i)^{\delta} \qquad \text{if } h \neq p $$

$$ {}_tR_p = {}_t\pi_p + {}_tR_{p-1} \cdot (1+i)^{\delta} $$

Thus, if $h \neq p$:

$$ {}_t\pi_h = {}_t\pi_{h-1} \cdot (1+i)^{\delta}, $$

$$ {}_t\gamma_h = {}_t\gamma_{h-1} \cdot (1+i)^{\delta}; \tag{22} $$

and if $h = p$:

$$ {}_t\pi_p = {}_t\pi_{p-1} \cdot (1+i)^{\delta} + {}_t\pi_p - {}_{t-1}\pi_p \cdot (1+i), $$

$$ {}_t\gamma_p = {}_t\gamma_{p-1} \cdot (1+i)^{\delta} + {}_t\pi_p - {}_{t-1}\pi_p. $$

The first of these two equations can be rewritten

$$ {}_t\pi_{p-1} = {}_{t-1}\pi_p \cdot (1+i)^{1-\delta}. $$

Thus, emerging profits π_h for different accounting periods vary only by a constant interest factor. This implies that the internal rate of return y, which is defined as the solution of

$$ {}_nS(\pi_h;y) = \sum_{t=1}^{n} {}_t\pi_h \cdot (1+y)^{n-t} = 0, $$

does not depend on which accounting period is chosen.

Similarly, adjusted gains γ_h for different accounting periods vary only by a constant interest factor.

Both statements refer to the time periods between the realization of profits. If reading the matrices π and γ row by row, there will be jumps at transactions p, i.e., when new profits do emerge.

Also, it must be remembered that i = constant has been assumed in this section.

4.9. *Amortization Reserves and Statutory Reserves*

Let C be a cashflow matrix which does not reflect contributions to Statutory reserves.

Let C^a and C^b represent other cashflow matrices which are identical with C, except that they include the following additional transactions:

Case a: Statutory reserves ${}_tV^s$ are set up at steps (t,z) and released at steps (t+1,1).

Case b: Statutory reserve increases ${}_tV^s - (1+{}_ti) \cdot {}_{t-1}V^s$ are reflected in steps (t,z).

Let

$$R = R(C), \quad R^a = R(C^a), \quad R^b = R(C^b)$$

denote the corresponding reserves. Applying the recursion formula (20) repeatedly, we obtain the development of the differences $(R^a\text{-}R)$ and $(R^b\text{-}R)$ as shown in the schedules below, in which ${}_0V^s = 0$ is assumed.

(h,t)	${}_tC^a_h - {}_tC_h$	${}_tC^b_h - {}_tC_h$	${}_tR^a_h - {}_tR_h$	${}_tR^b_h - {}_tR_h$
(1,z-1)	0	0	0	0
(1,z)	$-{}_1V^s$	$-{}_1V^s$	$-{}_1V^s$	$-{}_1V^s$
(2,1)	$+{}_1V^s$	0	0	$-{}_1V^s$
...				
(2,z-1)	0	0	0	$-{}_1V^s \cdot (1+i)$
(2,z)	$-{}_2V^s$	${}_1V^s(1+i) -{}_2V^s$	$-{}_2V^s$	$-{}_2V^s$
(3,1)	$+{}_2V^s$	0	0	$-{}_2V^s$

This result can be interpreted as follows:

Reserves R "ignore" Statutory reserves.

Reserves R^a "ignore" Statutory reserves too, except that the elements ${}_tR^a_z$ of column z are "net" of Statutory reserves.

Reserves ${}_tR^b_z$ are "net" of the corresponding Statutory reserves ${}_tV^s$, while reserves ${}_{t+1}R^b_h$ are reduced by Statutory reserves ${}_tV^s$

carried forward with interest.

We may apply the same process to either asset-shares A or any amortization reserve V simply by replacing the symbol R by either A or V.

Thus, the difference A - V remains unaffected, that is

$$A - V = A^a - V^a = A^b - V^b.$$

It follows from the definitions

$$A - V = S(\pi;i) = S(\gamma;0)$$

that emerging profits π and also adjusted gains γ are the same, whether or not we include additional transactions regarding Statutory reserve increases.

We may rephrase this statement by saying that profits π and adjusted gains γ are determined by amortization reserves "modulo" Statutory reserves.

Thus, it does not really matter whether Statutory reserve increases are reflected in the cashflow matrix C, except that we could not determine IRR, the internal rate of return, if Statutory reserves were ignored.

4.10. *Extension of Notation*

Here is a capsule version of the theory presented in this paper:

$$\begin{aligned} A &= S(B;i) &= R(C) \\ V &= S(B-\pi;i) &= R(C') \\ A-V &= S(\pi;i) &= S(\gamma;0) \end{aligned}$$

Function S represents the accumulation process defined in Chapters 2.3 and 4.5. Function R refers to a similar process which is defined by equation (18) in Chapter 4.4. To underline the dependence upon the interest rate i, it would be appropriate to

replace : $A = R(C)$
by either : $A = R(C;i)$
or : $A = SS(C;i)$.

The choice SS has some appeal because of its resemblance to ΣΣ.

As far as the APL implementation is concerned, the choice is between

consequent notation, which would call for SS(C;i), and

CPU time as well as numerical accuracy, which makes R(C) preferable.

Indices may be introduced as follows:

$$ {}_tA_m = {}_tS(B_m;i) = {}_{(t,m)}SS(C;i) = {}_tR_m(C). $$

The different placement of index m shall indicate that asset shares A after event (t,m) can be computed

either by accumulation S, extended from 1 to t along the column *vector* B_m,

or by accumulation SS, extended from events (1,1) to (t,m) row after row through the *matrix* C.

PART 5. NUMERICAL EXAMPLE

Summary: The reserve matrix concept is demonstrated, using the model given by Paquin.

Language: APL

Dictionary:

i	I = vector, or INT = matrix
u	U
C	C
A = R(C)	A ← R C
B = S(A;i)	B ← INT S̲ A

C'	$\underline{C}$
V = R(C')	V ← R $\underline{C}$
$\pi = \underline{S}(A\text{-}V;i)$	PI ← INT $\underline{S}$ A-V
$\gamma = \underline{S}(A\text{-}V;0)$	G ← 0 $\underline{S}$ A-V
${}_tV_m$	V[T;M] = scalar - element of matrix V
V_m	V[;M] = vector = column m of matrix V.

5.1. Description of Model

The following numerical investigation is based on the example given in the paper of Claude Paquin (TSA XXV). It refers to an endowment at age 85, issued at age 35. After 25 policy years all surviving policies are assumed to lapse. Interest rate assumed is 6%.

Enclosed is a series of computer printouts.

Printout 1 contains (a) three headlines, followed by
(b) the *benefit matrix*.

Headline

1 : Column number and (behind the decimal point) signals which control the input of realized profits.

2 : Codes which link benefits to probabilities.

3 : Time scale (0 = beginning of calendar year,
1/2 = middle " " "
1 = end " " " .)

The benefit matrix refers to the calendar year as accounting period. We distinguish between "early" death claims which occur in the first half of the policy year, and "late" death claims which occur in the second half. (We apply to both the interest factors $v^{1/2}$ -- Paquin uses $v^{1/4}$ and $v^{3/4}$; the difference is negligible.)

The benefit matrix shows positive amounts for income and negative ones for expenditures. The columns of the benefit matrix (as well as the columns of all the following printouts except Printout 2) refer to the following transactions:

Column

1 : Statutory mean reserves (released)
2 : "Late" death benefits
3 : Dividends
4 : Cash surrender values
5 : Non-amortizable expenses
6 : Gross premium ($19.79 annually)
7 : Amortizable expenses
8 : "Early" death benefits
9 : Statutory mean reserves (set up)

Thus, the transactions occuring during an accounting period are taken up in the same order we probably would choose in case of manual computations. Nevertheless, the order is arbitrarily chosen. In the foregoing case, the release of Statutory reserves is effected in the first step. Thus, all reserves we shall calculate are "net" of Statutory reserves, except in the last step, i.e., column 9. At the end of this investigation, we shall study the impact of a different order of transactions.

Although we consider only 25 policy years, each computer printout records 27 accounting periods. This comes about as follows: First, the 25 policy years reach into 26 different calendar years. Second, transactions in calendar year 26 contribute to book results which emerge in calendar year 27.

We have "filled up" the benefit matrix with zeros. That is, transactions which do not occur (either because no policy has yet been issued, or because all policies have already lapsed) are considered as transactions of amount zero.

In the same spirit, we may attach the value zero to all "reserves" before the first non-zero transaction.

Printout 2 contains a matrix of *probabilities* (times 1,000).

For example, assuming for the first policy year $q = 0.00077$ and $w = 0.2$, we have:

Position:

Line 1, Col. 5	Entering first policy year	1,000.000
Line 1, Col. 6	Early deaths = 1,000 × q/2	0.385
Line 1, Col. 7	Surviving first calendar year	999.615
Line 2, Col. 1	Entering second " "	999.615
Line 2, Col. 2	Late deaths = 1,000 × q/2	0.385
Line 2, Col. 3	Surviving	999.230
Line 2, Col. 4	Lapses = 1,000 × (1-q) × w	199.846
Line 2, Col. 5	Surviving first policy year	799.384

Printout 3 contains the *cash flow matrix C.* C is the product of benefits times the appropriate probabilities, for instance:

Position:

Line 2, Col. 1	\$1.21 × 0.999615	=	\$ 1.210
Line 2, Col. 2	-\$1,000 × 0.000385	=	-\$ 0.385
Line 2, Col. 6	\$19.79 × 0.799384	=	\$15.820

Printout 4 contains UC = U × C. U is the discount factor at rate 6% annually. Examples:

Position:

Line 2, Col. 1	$\$1.210 \times 1.06^{-1}$	=	\$ 1.141
Line 2, Col. 2	$-\$0.385 \times 1.06^{-1}$	=	-\$ 0.363
Line 2, Col. 6	$\$15.820 \times 1.06^{-3/2}$	=	\$14.496

The sum over the columns of UC is shown at the bottom of Printout 4. We may use them to determine the natural reserve premium:

PV of total gain	7.295
+PV of not amortizable expenses	15.415
	22.710
divided by PV of gross premiums	126.855

gives 0.179.
Thus, the natural reserve premium is \$19.79 less the "margin" of 17.9% or \$16.247.

Paquin shows (in Table 7) the same figure: \$11.87 + \$4.38 = \$16.25.

5.2. *Computer Programs*

The APL programs used are displayed in *Printouts 5 and 6.*

In APL, programs are called "functions". They may depend on one argument (shown to the right of the function name), and on a second argument (to the left). For instance, the main function is called

C̲V∆MAT

which may be read: C̲reate matrices of reserves V. The function has the matrix shown in Printout 1 as left argument, and the interest assumption as the right argument.

Auxiliary functions are:

C̲L	to create the matrix shown in Printout 2
C̄U	to create the discount factors u(t,h)
S̄HAPE	to accommodate abbreviated input, if interest rates vary by calendar year
C̲IRR	to calculate the internal rate of return by means of iterations, using NEWTON and FUNC as subroutines.

In program step [6] of the main function, the asset share matrix A is calculated, using formula (18) of part 4 which is implemented as function R:

$$V \leftarrow R\ C \leftrightarrow V = R(C)$$

In program steps [8,9,10] the amortization reserves V for GAAP, IRR and Statutory accounting, respectively, are produced. This time, function R is applied to the cashflow matrix $\underline{C} \leftrightarrow C'$ after adjustment for the release of profits. The adjustment is effected by the subroutines C̲∆GAAP, C̲∆IRR and C̲∆STAT, respectively.

Finally, we find in the lower half of *Printout 6* the S-operator:

$$V \leftarrow I\ S\ B \leftrightarrow V = S(B;i)$$

and its partial inverse

$$B \leftarrow I\ \underline{S}\ V \leftrightarrow B = \underline{S}(V;i) = V - i._{*}V.$$

Also, a simple demonstration of how to use the functions S and S̲ is displayed.

5.3. *Asset Share Accumulation A*

Printout 11 contains the matrix A of asset share accumulations. The last column to the right displays the asset shares ${}_tA_z$ on a calendar-year basis. The other columns show ${}_tA_h$, assuming that the accounting year ends after the transaction of column h.

Printout 12 displays asset share factors, that is A divided by the persistency factors L. As expected, the difference between the last two columns to the right is equal to the Statutory reserves as shown in column 9 of Printout 1.

From both Printouts 11 and 12 we draw the important conclusion that Statutory reserves have an impact on column 9 only. That is, the asset shares on columns 1 of each year are equal to the asset shares of columns 8 of the preceding year.

Thus, in order to ignore Statutory reserves, we could simply leave out columns 1 and 9. This would suffice in case of GAAP accounting, but would not permit including IRR accounting into the same calculation process.

Printout 13 shows the matrix of book results B. Column 9 matches the "expected Statutory earnings" in Paquin's Table 8 as follows:

	Column 9	Paquin
Year 1	-9.231	-9.236
Year 11	1.461	1.462
Year 21	1.022	1.024

Printout 14 displays G, the gains from operations. As expected, the totals for each column are equal to the figures shown in the last line of Printout 11.

5.4. *Statutory Accounting*

Printout 21 shows "Statutory amortization reserves" V∆STAT. These are calculated under the assumption that profits π, which are equal to book results B from Printout 13, column 9, emerge in column 8.

Printout 22 contains the corresponding reserve factors.

Printout 23 displays "emerging statutory profits" π. As expected, columns 9 of Printouts 13 and 23 are identical.

Printout 24 shows the gains γ from operation in the statutory accounting case. They are different from the gains G in Printout 14, except for column 9, for the following reasons:

(a) Gains G in Printout 14 ignore statutory reserves, except in column 9. Thus, G represents actual cash earnings before increase of statutory reserves.

(b) Gains γ in Printout 24 are net of increases of statutory reserves. Thereby, statutory reserves for accounting periods other than calendar years are developed by formulae (20) of part 4. Thus, statutory terminal reserve factors as shown in column 2 of Printout 22 may not exactly match published statutory terminal reserve factors.

Printout 25 displays the difference A - V∆STA between asset shares A and statutory amortization reserves. Thus, row t of Printout 25 represents the total of statutory gains which were reported up to the year t.

5.5. *GAAP Accounting*

Printouts 31 through 35 replace Printouts 21 through 25, if GAAP accounting is applied.

The reserve factors shown in *Printout 32* agree with the corresponding figures given in Table 7 of Paquin with the following accuracy:

Initial Reserves	Column 7	Paquin
Year 1	- 7.208	- 7.20
Year 11	134.093	134.09
Year 21	315.626	315.62

Terminal reserves	Column 2	$\times 1.06^{1/2}$	Paquin
Policy year 1	- 8.198	- 8.440	- 8.44
Policy year 11	135.038	139.030	139.04
Policy year 21	317.683	327.075	327.08

Mean reserves	Column 8	Paquin
Year 1	- 7.809	- 7.82
Year 11	136.550	136.56
Year 21	321.339	321.35

We note again that columns 8 and 9 in Printout 32 are different by the amount of Statutory mean reserve factors as shown in Printout 1, column 9. Thus, under the reserve matrix concept it is just a matter of index h to the right of ${}_tV_h$ whether we refer to

initial reserves
terminal reserves
mean reserves including statutory mean reserves, or
mean amortization reserves defined as the excess over mean statutory reserves.

We also note that the amortization reserves as shown in the reserve matrix include statutory reserves, except for the columns "between" setting-up and release of statutory reserves (that is, only column 9 in the particular order of transactions being chosen).

Released profits as shown in *Printout 33*, column 9, agree with "expected adjusted earnings" in Table 8 of Paquin to the following extend:

	Column 9	Paquin
Year 1	-0.215	-0.205
Year 11	0.545	0.544
Year 21	0.333	0.334

5.6. *IRR Accounting*

Printouts 41 through 45 correspond to Printouts 21 through 25 and 31 through 35.

Starting from the book results as shown in Printout 33, column 9, we find the internal rate of return (IRR) to be 14.095%. IRR profits are assumed to emerge in column 8.

Comparison of Printouts 32 and 42 reveals that IRR amortization reserves are slightly higher than GAAP amortization reserves. Of course, this statement is restricted to the particular case under study.

Comparison of Printouts 34 and 44 shows that gains γ are higher for GAAP in most calendar years.

Finally, comparison of Printouts 35 and 45 reveals that, from the second year on, the accumulated GAAP earnings are higher than accumulated IRR earnings. However, the difference is rather insignificant. It must be left to further investigations whether this interesting observation can be generalized.

5.7. *Variations of the Order of Transactions*

Printouts 51 through 53 reflect a session on the APL terminal. The following happens.

First, we recall the headlines of Printout 1 and execute CVMAT. To check that we use the same data as before, we ask for the margin and the IRR.

Then, we want a computational proof that the IRR does not depend on the particular accounting period, such as calendar year or policy year. That is the case, since the IRR from all columns [;h] of the B matrix (as shown in Printout 23) turns out to be the same, namely 14.095%.

At the bottom of Printout 51 we store the reserves defined by the order of transactions being chosen so far. We do that by adding a "0" to each label (say, A0 for asset shares).

Finally, we reorder the benefit matrix such that transaction 1 (reserve release) occurs now between transactions 8 and 9.

Thus, the order of transactions has changed as follows:

Transaction as defined in Printout 1	is now transaction number
2 through 8	1 through 7
1	8
9	9

In *Printout 52* we calculate CVΔMAT based on the new order. As expected, the margin and the IRR have not changed.

Later in Printout 52 we define a new function EQUAL which produces the following two answers:

First line : a "1" if the two expressions to the left and the right of EQUAL are identical, a "0" if not.

Second line : a "1" if the two expressions to the left and right of EQUAL are equal within a tolerance of 10^{-10} for computational rounding errors, a "0" if not.

As expected, we find that the asset shares A (based on the new order of transactions) are, within the tolerance limits, equal to asset shares A0 (based on the original order of transactions) less the Statutory reserves (from column 1, Printout 21) times the appropriate interest factor.

Finally, on *Printout 53* we demonstrate that emerging profits π and adjusted gains γ are not affected by the change of the order of transactions, except for the transaction 8 (new), which refers to the release of Statutory reserves. This exception has no significance.

To summarize, we have demonstrated computationally that we may change the order of transactions in such a way that amortization reserves V may include Statutory reserves, or may not. The latter is the case for columns "between" release of the old Statutory reserve and the input of the new Statutory reserves.

Which alternative we choose has no impact on emerging profits π or adjusted gains γ.

REVERSIONARY ANNUITIES AS APPLIED TO THE EVALUATION OF LAW AMENDMENT FACTORS

Lena Chang

Actuarial and Insurance Consultant
Chang & Cummings
6 Beacon Street
Boston, Massachusetts

INTRODUCTION

Due to recent proposals for changes in the benefit structure for Workmen's (Worker's) Compensation laws on both the federal and state legislative levels, the evaluation of law amendment factors due to such legislative proposals becomes an everyday encounter for the casualty actuary concerned with this line of insurance. An important and involved part of such evaluations deals with survivor benefit provisions for both fatal and permanent and total cases. The analysis of survivor beneficiary classes is traditionally done according to Fratello's method as illustrated by Exhibits II, IIA, IIB in his article on Workman's Compensation [1]. Fratello used a number of approximations in both the assignment and the evaluation of annuities because of the limited computing facilities available at the time.

This study is mainly directed toward improving the assignment and evaluation of the life annuities associated with each of

ISBN 0-12-394680-8

the beneficiary dependency classes. The use of a combination of multiple life, multiple decrement (death and remarriage) multiple life, and reversionary annuities enables one to evaluate each dependency class reflecting exactly when and who should be receiving compensation at their respective benefit levels according to the specification of the actual or proposed Workmen's Compensation law. The calculation of the aforementioned annuity values is done by use of generalized commutation functions constructed directly from life and remarriage tables. Performing the calculations with the aid of computer programs also facilitates the evaluation for cases where select and ultimate tables must be applied. This study should be useful because it blends the theory of multiple life and multiple decrement annuities with available modern computational technologies for solving a casualty actuarial problem. It should be noted that this method eliminates the need for building a multiple decrement table for evaluating multiple decrement annuities; therefore, it is a much more direct and realistic method than that currently used by casualty actuaries.

This study is divided into three parts for presentation:

I. The logistics underlying annuity assignments to appropriate dependency classes;

II. The implementation for computations;

III. An illustration.... Federal Employees' Compensation Act (FECA) [2] proposed benefit structure for fatal cases.

I. THE LOGISTICS UNDERLYING ANNUITY ASSIGNMENTS

(a) For dependency classes containing a single beneficiary ...the assignment is simply a whole or temporary life annuity decrement by the specified cause(s) for cessation of compensation being paid to said beneficiary. For example, the class of widow alone would be assigned $\tilde{a}_y'$,[1] which is decremented by death and remarriage (before age 60 under FECA), where y is the pension average age of widow.

(b) For classes containing more than one initial beneficiary...the assignment must take care of all possible disjoint pairs of status w,z where the union of the lives in w with the lives in z is the set of all original dependent lives. The joint life status z will receive compensation for the unexpired term of its eligibility period as soon as, and only when, the last survivor status w fails or a specific time limit has been reached. Thus a reversionary annuity $\tilde{a}_{\overline{w:\overline{m|}}|z:\overline{t|}}$ must be used to determine the commuted weeks payable under this situation. In the case where w involves no life, $\tilde{a}_{\overline{w:\overline{m|}}|z:\overline{t|}} = \tilde{a}_{z:\overline{t|}}$. We note that m and t may be large enough to take all lives in w and z, to the end of mortality table. The advantage of our method of assignment is that we may easily calculate the average weekly benefit according to the specific rate of compensation due to the joint life of z. To illustrate, consider the initial beneficiary class of widow with one child: Let x = arithmetic average age of dependent children, y = pension average age of widow, n = the legal maximum age[2] for which the child beneficiary is eligible to receive benefits. Also assume that life y is decremented by remarriage.

Three disjoint situations may occur:

(1) both widow and child remain as beneficiaries, namely w involves no life........the assignment is simply $\tilde{a}_z'$ where z is the joint life $xy : \overline{n-x|}$.

(2) Child alone remains as beneficiary, namely $\overline{w} = y$, and $z = x$, $t = n-x$........the assignment is $\tilde{a}'_{y|x:\overline{n-x|}}$.

[1] $\tilde{a}$ *represents* $52a^{(52)}$*, for simplicity of notation, see definition in II(B)(1)(e).*

[2] *This may vary, e.g., if child is disabled or in school.*

(3) Widow alone remains as beneficiary, namely w = x and z = y.......the assignment is $\tilde{a}'_{x:\overline{n-x}|}|y$.

II. THE IMPLEMENTATION FOR COMPUTATION

(A). *Generalized Commutation Functions*

Given a single decrement (death, remarriage, or disability, etc.) life function ℓ_x, we define for each integer $n \geq 0$,

$$D_x^{(n*)} = \left(\frac{1+j}{1+i}\right)^{\overline{\frac{x}{n+1}}} \ell_x$$

where i = interest rate, j = the benefit escalation rate. From this point on, we shall fix the following notations:

ℓ_x from U.S. Standard Ordinary life table (1960) for total population;

ℓ'_x from Standard Ordinary life table (1960) for female population (U.S.);

ℓ^r_x from Remarriage table (ultimate) U.S. Employees' Compensation System; and

ℓ''_x from Disabled life table (ultimate) (whichevery applies).

The The corresponding $D_x^{(n*)}$'s will have corresponding superscript designations.

(B). *Annuities Certain, Ordinary and Term Life Annuities, Multiple Life Annuities, Reversionary Annuities, Double Decrement Multiple Life Annuities*

(1) All annuity values are calculated according to the following specifications:

(a) \$1 payable weekly,

(b) average incurrence date of all cases being mid year,

(c) future benefits discounted to present value at an effective annual interest rate i,

(d) future benefits escalated on anniversary according to annual benefit escalation rate j,

(e) symbols used for these annuities are $\tilde{a}$, or $\tilde{a}'$ in place of the conventional $52a^{(52)}$ and $52a'^{(52)}$ for tabulation purposes.

(2) Annuity certain for n-years,

$$\tilde{a}_{\overline{n}|} = v^{\frac{1}{2}} \times 52 \times (i/i^{(52)}) \times a_{\overline{n}|}$$

where

$$a_{\overline{n}|} = \text{annuity value of \$1 per year for n-years, payable at the end of each year escalated starting with the 2nd payment}$$

$$= v + (1+j)v^2 + \ldots + (1+j)^{n-1}v^n$$

$$= v\left[\sum_{t=0}^{n-1} \left(\frac{1+j}{1+i}\right)^t\right]$$

where

$$v = \frac{1}{1+i}$$

$$i^{(52)} = \text{nominal rate of interest payable 52 times a year.}$$

(3) Temporary life annuities and ordinary whole life annuities: Let z represent arithmetics average age of the other dependents, y represents the arithmetic average age of widow, x represents the average arithmetic age of children dependents, m = n-x the maximum period of compensation payable to x. Then,

$$\tilde{a}_{x:\overline{m}|} = v^{\frac{1}{2}} \times 52 \times \left[a_{x:\overline{m}|} + \frac{51}{104}\left(1 - \frac{D_{x+m}}{D_x}\right)\right]$$

$$\tilde{a}_x = v^{\frac{1}{2}} \times 52 \times \left[a_x + \frac{51}{104}\right]$$

where

$$a_{x:\overline{m}|} = v \cdot \frac{\sum_{t=1}^{m} D_{x+t}}{D_x}$$

and

$$a_x = a_{x:\overline{p}|},\ p = 109 - x$$

$$\tilde{a}'_{y:\overline{n}|} = v^{\frac{1}{2}} \times 52 \times \left[a'_{y:\overline{n}|} + \frac{51}{104}\left(1 - \frac{D_{y+n}}{D_y}\right)\right]$$

$$a'_{y:\overline{n}|} = v \cdot \frac{\sum_{t=1}^{n} D'^{*}_{y+t} \; D^{r*}_{y+x}}{D'^{*}_{y} \; D^{r*}_{y}}$$

where prime in general denote annuities decremented by both death and remarriage, n an integer.

(4) Multiple life annuities involving the life y, e.g.

$$\tilde{a}'_{xy:\overline{n}|} = v^{\frac{1}{2}} \times (52) \times (1+i)^{\frac{1}{2}} \times (v) \cdot \frac{\sum_{t=1}^{n} D^{**}_{x+t} D'^{**}_{y+t} D^{r**}_{y+t}}{D^{**}_{x} \; D'^{**}_{y} \; D^{r**}_{y}}$$

$$\tilde{a}'_{xy} = \tilde{a}'_{xy:\overline{n}|}, \quad n = \min\{109 - y,\ 109 - x\}$$

$$\tilde{a}'_{xyz:\overline{n}|} = v^{\frac{1}{2}} \times (52) \times (1+i)^{\frac{1}{2}} \times (v) \times \frac{\sum_{t=1}^{n} D^{***}_{x+t} D'^{***}_{y+t} D^{r***}_{y+t} D^{***}_{z+t}}{D^{***}_{x} \; D'^{***}_{y} \; D^{r***}_{y} \; D^{***}_{z}}$$

$$\tilde{a}'_{xyz} = \tilde{a}'_{xyz:\overline{n}|} \quad \text{where} \quad n = \min\{109 - x,\ 109 - y,\ 109 - x\}$$

where $D^{(n*)}_{x}$'s are as defined in (A).

(5) Reversionary annuities,

$$\tilde{a}'_{x:\overline{n}|\,|y} = \tilde{a}'_{y} - \tilde{a}_{x:\overline{n}|}$$

$$\tilde{a}'_{x|y} = \tilde{a}'_{y} - \tilde{a}'_{xy}$$

$$\tilde{a}'_{y|x:\overline{n}|} = \tilde{a}_{x:\overline{n}|} - \tilde{a}'_{xy:\overline{n}|}$$

(c). The Implementation for Computer Calculation

The formulae used for annuities listed in (B) can be directly used in a computer program with subroutines (SUMPRD). In the program, for each interest rate i, and benefit escalation rate j, we calculate complete tables for $D^{(n*)}_{x}$, $D'^{(n*)}_{x}$, $D^{r(n*)}_{x}$ and

$D''^{(n*)}_x$, n = 0,1,2,3 and ω = 109. Using these tables, the program calculates all the annuity values needed, including the approximations to them which are payable 52 times a year.

To take care of the t-year select period of remarriage, for each life age y_o using ℓ^r_y (the ultimate remarriage decrement table), we feed in t select values of $\ell^r_{[y_o]}, \ell^r_{[y_o]+1}, \ldots, \ell^r_{[y_o]+t-1}$ replacing $\ell^r_{y_o}, \ell^r_{y_o+1}, \ldots, \ell^r_{y_o+t-1}$ in the ultimate remarriage table. Consequently, the corresponding t values for $D^{r(n\ )}_{y_o}$, $D^{r(n*)}_{y_o+t-1}$ are replaced by $D^{r(n*)}_{[y_o]}, \ldots, D^{r(n*)}_{[y_o]+t-1}$. Thus all annuities involving the life y_o, are calculated according to the select and ultimate values of remarriage tables.

If there is a select period for disabled lives, a similar method can be applied on their select period.

III. AN ILLUSTRATION. FECA PROPOSED BENEFIT LEVEL CHANGE FOR FATAL SURVIVOR TO MEET THE NATIONAL COMMISSION ON STATE WORKMEN'S COMPENSATION LAWS LEVEL

Using the computer printout of annuity values, and formulae in (B) (5) for reversionary annuities one can easily obtain values for the entries in column (6) corresponding to each entry of column (5) of Table A, B. For example, for the dependency class of widow with one child (the 2nd, 3rd, 4th rows) according to (I), three disjoint situations arise. Column (3) describes the person(s) receiving compensation in each situation, column (4), gives the average age[1] of the beneficiaries, in column (5)

[1]*The child's average age is taken to be 10 instead of 8 as in Fratello to reflect the FECA law of continuation of compensation to child dependents as long as they are students up to 4 years beyond high school education. This is also the reason for the use of age 19 as the maximum age for compensation payable to child beneficiary.*

TABLE A FATAL Benefit Provision * (Excluding Remarriage Dowry)

VALUATION of FECA as amended 1974

Remarriage before age 60 decremented (Select-ultimate remarriage table U.S.E.C. System).
Absolute minimum benefit = 0, Minimum benefit = $81.95 = .75 X GS2
Maximum Benefit = $407.64 = .75 X GS15.
Benefits payable weekly, approximation on continuous annuity are assumed to start in the middle of the year and escalated by CPI index or anniversary.

(1) No. of cases	(2) No. of dependents	(3) prsn(s) rec'ing comp.	(4) average age	(5) annuity symbols
342	1	widow alone	50	$\bar{a}'_{50}$
155	2	wid, 1ch	35, 10	$\bar{a}'_{10:35:\overline{9}\rceil}$
155	1	child	10	$\bar{a}'_{35\vert 10:\overline{9}\rceil}$
155	1	widow	35	$\bar{a}'_{\overline{10:\overline{9}\rceil}\vert 35}$
117	3	wid, 2ch	35,10,10	$\bar{a}'_{10:10:35:\overline{9}\rceil}$
117	2	2 child	10, 10	$\bar{a}'_{35\vert 10:10:\overline{9}\rceil}$
117	2	wid, 1ch	35, 10	$2 \times \bar{a}'_{10\vert 35:10:\overline{9}\rceil}$
117	1	1 child	10	$2 \times \bar{a}'_{\overline{10:35}\vert 10:\overline{9}\rceil}$
117	1	widow	35	$\bar{a}'_{\overline{10:10:\overline{9}\rceil}\vert 35}$
124	4 or more	wid, 3ch+	35, 10 10, 10	$\bar{a}'_{\overline{10:10:10:35:\overline{9}\rceil}}$
124	3 n= #of ch	wid, 2ch	35,10,10	$\binom{n}{2}\bar{a}'_{\overline{10}\vert 10:10:35:\overline{9}\rceil}$

(This calculation was originally done for costing of the proposed benefit changes under FECA contract #J-9-E-5-0050.)

Benefits escalated annually according to CPI index $j = .06$.
Present value calculated at interest rate $i = .04$.
Injured Workers' AWW Ratio adjusted by country wide experience = \$212.34 x 1.06 = \$225.08, escalaued to 1975.

(6) annuity values \$1 per wk	(7) Rate of comp.	(8) average wkly wage 1974	(9) average wkly benefit 1975	(10) Cost Excluding remarriage dowry (1)x(6)x(9)
1835.46	50%	241.65	122.82 x 1.06	81,723,317
416.25	60%	241.65	133.25 x 1.06	9,112,951
63.57	40%	241.65	114.30 x 1.06	1,193,812
1952.83	50%	241.65	122.82 x 1.06	39,406,793
403.53	75%	241.65	159.97 x 1.06	8,005,825
88.58	55%	241.65	125.55 x 1.06	1,379,254
2 x 12.72	60%	241.65	133.25 x 1.06	420,413
2 x 0	40%	241.65	114.30 x 1.06	0
1965.56	50%	241.65	122.82 x 1.06	29,939,678
403.53	75%	241.65	159.97 x 1.06	8,484,806
0	75%	241.65	159.97 x 1.06	0

TABLE A (continued)

(1) No. of cases	(2) No. of dependents	(3) prsn(s) rec'ing comp.	(4) average age	(5) annuity symbols
124	2 n= # of ch	wid, 1ch	35, 10	$\binom{n}{1} \times \bar{a}'_{\overline{10:10}\|35:10:\overline{9}\rceil}$
124	2	2 child	10, 10	$\binom{n}{2} \times \bar{a}'_{\overline{10:35}\|10:10:\overline{9}\rceil}$
124	1	1 child	10	$\bar{a}'_{\overline{10:10:35}\|10:\overline{9}\rceil}$
124	3 or more	3 child	10,10,10	$\bar{a}'_{35\|10:10:10:\overline{9}\rceil}$
124	1	widow	35	$\bar{a}'_{\overline{10:10:11}:9:35}$
18	1	orphan	11	$\bar{a}_{11:\overline{8}\rceil}$
10	2	2 orph	11, 11	$\bar{a}_{11:11:\overline{8}\rceil}$
10	1	1 orph	11	$2 \times \bar{a}_{11\|11:\overline{8}\rceil}$
8	3 or more	3 orph	11,11,11	$\bar{a}_{\overline{11}:11:11:\overline{8}\rceil}$
8	2 n =	2 orph	11, 11	$\binom{n}{2} \bar{a}_{\overline{11}\|11:11:\overline{8}\rceil}$
8	1 # of ch	1 orph	11	$\binom{n}{1} \bar{a}_{\overline{11:11}\|11:\overline{8}\rceil}$
40	1	parent	61	$\bar{a}_{61}$
27	2	2 par	56, 56	$\bar{a}_{56:56}$
27	1	1 par	56	$2 \times \bar{a}_{56\|56}$
4	1	sister or brother	43	$\bar{a}_{43}$
11	2 or more	other dep	43, 43	$\bar{a}_{\overline{43}:43}$
11	1 n=#of dep	other dep	43	$\binom{n}{1} \bar{a}_{\overline{43}\|43}$
5 { $\frac{1}{4}$	2	wid & 1 dep	50, 43; 50, 61	$\bar{a}'_{50:43}$ $\bar{a}'_{50:61}$
5 { $\frac{1}{4}$	1	1 dep	43; 61	$\bar{a}'_{50\|43}$ $\bar{a}'_{50\|61}$
5 { $\frac{1}{4}$	1	widow	50	$\bar{a}'_{43\|50}$ $\bar{a}'_{61\|50}$
139	0	none		
1000				

(6) annuity values $1 per wk	(7) Rate of comp.	(8) average wkly wage 1974	(9) average wkly benefit 1975	(10) Cost Excluding remarriage dowry (1)x(6)x(9)
0	60%	241.65	133.25 x 1.06	0
0	55%	241.65	125.55 x 1.06	0
0	40%	241.65	114.30 x 1.06	0
86.87	55%-75%	241.65	(125.55-159.97) x 1.06	1,826,568
1963.99	50%	241.65	122.82 x 1.06	31,705,596
422.37	40%	241.65	114.30 x 1.06	921,123
433.18	55%	241.65	125.55 x 1.06	576,489
2 x 0	40%	241.65	114.30 x 1.06	0
430.17	70%-75%	241.65	(151.41-159.97) x 1.06	583,545
0	55%	241.65	125.55 x 1.06	0
0	40%	241.65	114.30 x 1.06	0
929.26	25%	241.65	111.42 x 1.06	4,390,018
754.96	40%	241.65	114.30 x 1.06	2,469,675
2 x 416.51	25%	241.65	111.42 x 1.06	2,656,368
2018.43	20%	241.65	108.38 x 1.06	927,532
1428.44	30%	241.65	113.47 x 1.06	1,889,912
590	20%	241.65	108.38 x 1.06	745,589
1374.84 780.38	70%-75%	241.65	(151.41-159.97) x 1.06	749,965
539.98 → 148.88	20%-25%	241.65	(108.38-111.42) x 1.06	132,368
460.62 →1055.08	50%	241.65	122.82 x 1.06	609,408
	0%	241.65		
				229, 851, 005

TABLE B FATAL Benefit Provision * (Excluding Remarriage Dowry)

Remarriage before age 60 decremented (Select-ultimate remarriage table U.S.E.C. System).
Absolute minimun benefit = 0, Minimun benefit = .5 x 241.65 = 120.82, Maximum benefit = 2 x 241.65 = 483.3
Benefits payable weekly, approximation on continuous annuity are assumed to start in the middle of the year and escalated by CPI index on anniversary.

(1) No. of cases	(2) No. of dependents	(3) prsn(s) recv'ing comp.	(4) average age	(5) annuity symbols	(6) annuity values $1 per wk
342	1	wid. alone	50	$\bar{a}'_{50}$	1835.46
155	2	wid, 1 ch	35, 10	$\bar{a}'_{10:35:\overline{9}\rceil}$	416.25
155	1	child	10	$\bar{a}'_{35\|10:\overline{9}\rceil}$	63.57
155	1	widow	35	$\bar{a}'_{10:\overline{9}\rceil\|35}$	1952.83
117	3	wid, 2ch	35,10,10	$\bar{a}'_{10:10:35:\overline{9}\rceil}$	403.53
117	2	2 child	10, 10	$\bar{a}'_{35\|10:10:\overline{9}\rceil}$	88.58
117	2	wid, 1 ch	35, 10	$2 \times \bar{a}'_{10\|10:35:\overline{9}\rceil}$	2 x 12.72
117	1	1 child	10	$2 \times \bar{a}'_{\overline{10:35}\|10:\overline{9}\rceil}$	2 x 0
117	1	widow	35	$\bar{a}'_{\overline{10:10}:\overline{9}\rceil\|35}$	1965.56
124	4	wid, 3ch+	35, 10 10, 10	$\bar{a}'_{10:10:10:35:\overline{9}\rceil}$	403.53
124	3 n= # ofch	wid, 2ch	35,10,10	$\binom{n}{2} \times \bar{a}'_{\overline{10}\|35:10:10:\overline{9}\rceil}$	0
124	2	wid, 1ch	35,10	$\binom{n}{1} \times \bar{a}'_{\overline{10:10}\|35:10:\overline{9}\rceil}$	0

(This calculation was ariginally done for costing of the proposed benefit changes under FECA contract #J-9-E-5-0050.)

Benefits escalated annually according to CPI index j = .06.
Present value calculated at interest rate i = .04.

(7) Rate of comp.	(8) average weekly wage 1974	(8') spendable weekly wage 1974	(9) average weekly benefit 1974	(10) Cost Excluding remarriage dowry (1)x(6)x(9) x (1.06)
80%	241.65	184.96	154.71	102,942,635
80%	241.65	188.20	157.42	10,765,933
80%	241.65	184.96	154.71	1,615,876
80%	241.65	184.96	154.71	49,638,699
80%	241.65	191.21	159.94	8,004,324
80%	241.65	188.20	157.42	1,729,368
80%	241.65	188.20	157.42	496,671
80%	241.65	184.96	154.71	0
80%	241.65	184.96	154.71	37,713,464
80%	241.65	194.26+	161.24	8,552,167
80%	241.65	191.21	159.94	0
80%	241.65	188.20	157.42	0

TABLE B (continued)

(1) No. of cases	(2) No. of dependents	(3) prsn(s) recv'ing comp.	(4) average age	(5) annuity symbols	(6) annuity values $1 per wk
124	2	2 child	10,10	$\binom{n}{2}\ \bar{a}'_{\overline{10:35}\mid 10:10:\overline{9}\rceil}$	0
124	1	1 child	10	$\bar{a}'_{\overline{10:10:35}10:\overline{9}\rceil}$	0
124	3 or more	3 child	10,10,10	$\bar{a}'_{35:10:10:10:\overline{9}\rceil}$	86.87
124	1	widow	35	$\bar{a}'_{\overline{10:10:10}:\overline{9}\rceil\mid 35}$	1963.99
18	1	orphan	11	$\bar{a}_{11:\overline{8}\rceil}$	422.37
10	2	2 orph	11, 11	$\bar{a}_{11:11:\overline{8}\rceil}$	433.18
10	1	1 orph	11	$2 \times \bar{a}_{11\mid 11:\overline{8}\rceil}$	2 x 0
8	3 or more	3 orph	11,11,11	$\bar{a}_{11:11:11:\overline{8}\rceil}$	430.17
8	2 n= #of dep	2 orph	11, 11	$\binom{n}{2}\times\bar{a}_{11\mid 11:11:\overline{8}\rceil}$	0
8	1	1 orph	11	$\binom{n}{1}\times\bar{a}_{\overline{11:11}\mid 11:\overline{8}\rceil}$	0
40	1	parent	61	$\bar{a}_{61}$	929.26
27	2	2 parent	56, 56	$\bar{a}_{56:56}$	754.96
27	1	1 parent	56	$\bar{a}_{56\mid 56}$	2 x 416.51
4	1	sister or brother	43	$\bar{a}_{43}$	2018.43
11	2 or more	other dep	43, 43	$\bar{a}_{\overline{43}:43}$	1428.44
11	1 n= #of dep	other dep	43	$\binom{n}{1}\times\bar{a}_{43\mid 43}$	590
5 $\{^{1}_{4}$	2	wid, 1dep	50, 43; 50, 61	$\bar{a}'_{50:43}$ → $\bar{a}'_{50:61}$	1374.84 →780.38
5 $\{^{1}_{4}$	1	1 dep	43, 61	$\bar{a}'_{50\mid 43}$ → $\bar{a}'_{50\mid 61}$	539.98 →148.88
5 $\{^{1}_{4}$	1	widow	50	$\bar{a}'_{43\mid 50}$ → $\bar{a}_{61\mid 50}$	460.62 →1055.08
139	0	none			
1000					

(7) Rate of comp.	(8) average weekly wage 1974	(8') spendable weekly wage 1974	(9) average weekly benefit 1974	(10) Cost Excluding remarriage dowry (1)x(6)x(9) x (1.06)
80%	241.65	188.20	157.42	0
80%	241.65	184.96	154.71	0
80%	241.65	191.21+	159.94+	1,826,226
80%	241.65	184.96	**154.71**	47,746,839
80%	241.65	184.96	154.71	1,246,780
80%	241.65	188.20	157.42	722,827
80%	241.65	184.96	154.71	0
80%	241.65	191.21+	159.94+	583,436
805	241.65	188.20	157.42	0
80%	241.65	184.96	154.71	0
80%	241.65	184.96	154.71	6.095,670
80%	241.65	188.20	157.42	3,401,367
805	241.65	184.96	154.71	3,688,446
80%	241.65	184.96	154.71	1,324,030
80%	241.65	188.20+	157.42+	2,621,926
80%	241.65	184.96	154.71	1,064,312
80%	241.65	188.20	157.42	750,286
80%	241.65	184.96	154.71	186,213
80%	241.65	184.96	154.71	767,640
80%	241.65	241.65	0	
				293,485,127

row 2, 3, 4 are the annuity assignments according to (I). Corresponding to these, column (6), rows 2, 3, 4 are calculated as following:

$$\text{(row 2)} \quad \tilde{a}_{10:35:\overline{9}|} = 416.25 \text{ weeks}$$

$$\text{(row 3)} \quad \tilde{a}_{35|10:\overline{9}|} = \tilde{a}_{10:\overline{9}|} - \tilde{a}'_{10:35:\overline{9}|}$$

$$= 479.82 - 416.25 = 63.57 \text{ weeks}$$

$$\text{(row 4)} \quad \tilde{a}'_{10:\overline{9}||35} = \tilde{a}'_{35} - \tilde{a}'_{10:35:\overline{9}|}$$

$$= 2369.08 - 416.25 = 1952.83.$$

Column 7 gives the corresponding rate of compensation specified by law for each situation described in column 5 and column 9 gives the average weekly benefit calculated according to the rate of compensation specified by column 7.

Tables A and B illustrate the calculation for the cost of survivor provision per 1000 fatal cases on incurred basis. The cases are distributed according to Fratello's dependency distribution. The calculations were made using the method discussed in this paper. The evaluation is based upon the Federal Employee's Compensation Act (FECA) existing benefit structure (A) and the proposed Benefit structure (B) -- namely that recommended by the NCSWCL.

The percent increase in cost due to the change in benefit structure for fatal survivor benefits without dowry =

$$\left(\frac{293,485,127}{229,851,005} - 1\right) \times 100\% = 27.6\%.$$

REFERENCES

1. "Federal Employee's Compensation Act, as Amended", Including changes made by Public Law 93-416, approved September 7, 1974, Prepared by the Subcommittee on Labor of the Committee on Labor and Public Welfare United States Senate, October 1974. U.S. Government Printing Office.
2. Fratello, Barney. "Workmen's Compensation Injury Table" and "Standard Wage Distribution Table", Reprinted from the Proceedings of the Casualty Actuarial Society Volume XLII.
3. "The Report of the National Commission on State Workmen's Compensation Laws", Library of Congress Card Catalog No. 72-600195, July 1972, pps. 60-62.

NONLIFE BUSINESS AND INFLATION

Harald Bohman

Skandia Insurance Company Ltd.
Box S-103 60
Stockholm, Sweden

In insurance business it is of vital importance to be able to calculate *present values*. The premium is the present value of future claim payments and costs of administration. The reserve for outstanding claims is the present value of future payments upon claims already incurred but still not settled. The premium reserve, often known as the unearned premiums, is in practice calculated from the premiums earned on a pro rata temporis basis, but is in fact the present value of expected costs for future claims upon insurfance contracts in force at the time of the valuation.

Since a present value has to do with the future and we do not know about the future we have to make assumptions. These assumptions are often known as the *technical base* and consist of assumptions regarding claim frequency, mean claim amount, mean duration from occurrence of claim until final settlement of claim, costs of administration, interest and inflation. The assumptions just mentioned are not a complete list of assumptions but just an example of assumptions which must be found in the technical base. In the present context we are primarily interested in the prob-

ISBN 0-12-394680-8

lems related to interest and inflation.

The need for an assumption regarding expected future inflation is characteristic of non-life insurance business. In life business it is traditional not to offer any value-preserving clause in the contract. Consequently the terms of the contract are fulfilled if the insurance company can pay to the policyholders the nominal values as set out in the contracts. It follows that the insurance company can forget about inflation as far as payments to the policyholders are concerned. As far as his costs of administration are concerned they are, however, subject to increases due to inflation and consequently the life companies ought to include in their technical base an assumption regarding future inflation to be applied to the administrative costs. It seems, however, that this is not done in practice.

Coming back to non-life business we recall that non-life insurance contracts generally aim at compensation for losses incurred. This compensation should be paid in the monetary value valid at the time of the occurrence of the claim, thus enabling the policyholders to replace or repair the material loss he has met with. Consequently we must not forget in non-life business that we are dealing with real values and that we need an assumption regarding future inflation in our technical base.

It is well-known that inflation can be measured in different ways. This has primarily not to do with different statistical and economic definitions of inflation but reflects the fact that during times of inflation different sectors of trade and industry seem to experience different rates of price increases. This situation also applies to the insurance industry, and consequently we have to face the fact that the rate of inflation might be different in different lines of business. For the sake of simplicity we will, however, disregard this fact in the sequel and treat the insurance business as if one and the same rate of inflation applies to all lines of business.

It is well-known that interest and inflation work in opposite direction and that you can in practice consider the difference between the rate of interest and the rate of inflation. This difference will be denoted by the letter r and will be our assumed *rate of real interest*. Having made our assumption regarding the rate of real interest r we have to apply it to all calculations of premiums. Our premiums could then be described as present values of future claim payments and administrative costs, where the present values are calculated on the assumption of a real interest r.

The insurance fund is here meant to be the sum of the reserve for outstanding claims and the premium reserve. As was described above the insurance fund is basically a present value in the same way as the premium is a present value. In order to make sure that we are using a consistent system we must apply the same technical base for the valuation of the insurance fund as is used for the premium calculation. This means for example that if we find that our premium rates are too low we have to strengthen our technical base and as a result our premium rates and our insurance fund will be strengthened.

A premium paid by a policyholder to his insurance company means that the insurance company has assumed a new responsibility towards this client. The present value of this responsibility is equal to the premium just paid by the policyholder and since the insurance fund shall represent an estimate of the liability of the company towards its policyholders we should add the premium paid to the insurance fund.

That the premiums paid to the insurance company should be added to the insurance fund seems quite clear from logical reasons. A much more difficult problem is to decide upon the rate according to which money might be taken out from the insurance fund in order to meet claim payments and administrative costs. For different reasons, including similarities between insurance business and the delay mechanisms often used in operational analysis, it seems reasonable to talk of a certain *rate of decrease*. If the insurance fund is denoted by F and the rate of decrease by q the product qF represents the *expected* payments for outstanding claims and administrative costs.

We now come to the problem how much interest we shall bring into the insurance fund. We denote by j^* the present rate of actual inflation. We then find that we have to increase the insurance fund by interest and this interest has to be calculated at the rate $(j^* + r)$.

What has been said above can now be expressed in the form of a differential equation for F as follows

$$\frac{dF}{dt} = (j^* + r)F + P - qF.$$

Assuming that premiums are received at a constant rate during the year and that j^* also remains constant during the year we can integrate this differential equation from the beginning of one year to the end of the same year, arriving at the following solution

F_o = insurance fund at beginning of year

F_1 = insurance fund at end of year

$$F_1 = AP + BF_o$$

$$B = \exp(j^* + r - q)$$

$$A = (B - 1)/(j^* + r - q)$$

It is seen from the differential equation how premiums and "interest" are fed into the insurance fund, while qF is taken out and we called this quantity "expected payments". Let us denote the corresponding actual payments by U. We find that according to our technical system the quantity qF is made available for payments, while U are the actual payments. The difference between these quantities will be our "underwriting profit", that is our profit from insurance business.

As long as $(qF - U)$ remains positive we are on the safe side. The amounts made available by our technical system for payments are sufficient to cover the actual payments. If, however, the opposite situation occurs we are making an underwriting loss. This loss has to be covered from the surplus fund and we have to take action in order to prevent further losses. The actions to be taken are a suitable increase of the premium income P and a suitable increase of the insurance fund F.

As for the premium increase we observe that the quotient U/qF tells us in what proportion we should increase our premium income P and our insurance fund F in order to put our system in a state of balance, where the expected and actual payments are the same. If this quotient is larger than 1 we then take the decision to increase our premium income in that proportion.

As for the strengthening of F we choose, however, a different method. We had at the beginning of the year an insurance fund equal to F_o and a premium income during the year equal to P. The amount of money made available for actual payments was found to be insufficient and resulted in an underwriting loss W. Is it possible to choose an initial value of F at the beginning of the year in such a way that the underwriting profit becomes equal to 0. This initial value of F will evidently be larger than F_o and the corresponding value of F at the end of the year will be larger than F_1. Our intention now is to derive an expression for the increase to be made in F_1.

Our rate of decrease is q and an approximate expression for the amount of money made available during one year is obtained by multiplying the mean value of the insurance fund by q. We then get the following equation since we had an underwriting loss W.

$$q(F_o + F_1)/2 = U - W$$

If the insurance fund had been instead F'_o and F'_1 at the beginning and the end of the year respectively we obtain if we choose F'_o

in such a way that no underwriting loss occurs

$$q(F_o' + F_1')/2 = U.$$

Subtracting these two equations after having expressed F_o and F_o' as functions of F_1 and F_1' we arrive at the following expression

$$F_1' - F_1 = \frac{2Bw}{q(1 + B)}$$

From this equation we now draw the conclusion that we have to strengthen the insurance fund at the end of the year by the amount as calculated according to the right-hand side of the above equation. Note that this strengthening takes place at the end of the year. Note also that our rule of decision says that if there is no underwriting loss there is also no need to strengthen the insurance fund. This need arises only when there is an underwriting loss, that is when W is positive.

Nothing has been said until now about the numerical value of q. It is seen from the differential equation for F that F will be larger the smaller q is. The size of q depends upon the mean duration from premium payment to payments for claims or administrative costs. This mean duration varies between lines of business. A value of q equal to 0.4 might be expected for accident and motor third party liability. For motor other than third party liability a value of q equal to 1.5 has been observed. Other observations are fire with q = 0.6 and marine with q = 0.8. Generally speaking it can be said that q should be chosen with respect to the experience within each particular line of business.

Summarizing what has been said above we can say that the intention is to describe a plan for a non-life insurance business, where the technical problems related to premium assessment and the assessment of insurance fund are treated in a consistent way and where special emphasis has been put upon inflation and interest. Looked upon from a traditional actuarial point of view it might seem somewhat revolutionary to introduce a differential equation governing the development of the insurance fund in the way described above. This is, however, necessary in order to get a mathematical treatment of the problem. There is no intention to recommend such a method to replace the well-known per case assessment of each individual outstanding claim. For the purpose of describing and analyzing how inflation affects our business the method is, however, useful. Practical experiments made by the author encourage further experiments with this model, where a possible application of the model is as a checking device whereby the accuracy of the insurance fund might be tested.

NUMERICAL FOURIER INVERSION

Harald Bohman

Skandia Insurance Company Ltd.
Box S-103 60
Stockholm, Sweden

Much has been written about the numerical inversion of characteristic functions, using more or less sophisticated methods. It is important, however, to notice that the real step forward in this area was taken in 1965 when the Fast Fourier Transform (FFT) was presented. The important thing is to be able to manipulate Fourier series with a large number of terms. If you can do that and it is possible with the aid of FFT, then you can get remarkably good results using even unsophisticated methods. In order to get real good results it might be necessary to introduce more sophisticated methods but the major improvement in accuracy is obtained from FFT.

An unsophisticated approach to the problem might be the following. Starting with a characteristic function $\phi(t)$ and a positive quantity T we define a periodic function ϕ_1, as follows

$$\phi_1(t) = \phi(t) \qquad |t| < T$$

$$\phi_1(T) = (\phi(T) + \phi(-T))/2$$

$$\phi_1(t + 2T) = \phi_1(t).$$

ISBN 0-12-394680-8

ϕ_1 is periodic with period 2T and it might be tempting to try to represent it by a Fourier series as follows

$$\phi_1(t) = \sum_{-\infty}^{+\infty} p_n \exp(\pi\, int/T).$$

As is seen from this formula ϕ_1 is expressed in the form of a characteristic function representing a discrete probability distribution where the different masses of probability are represented by the coefficients p_n. We now define the following quantities

$$d = 2\pi/N$$
$$T = \pi/d$$
$$\omega = \exp(2\pi i/N).$$

Our next step is to approximate ϕ_1 by a trigonometric polynomial of degree (N - 1). In order to do that we want the following N equations to be fulfilled

$$\phi_1(m\delta) = \sum_{0}^{N-1} p_n \omega^{mn}, \qquad m = 0(1)N - 1.$$

It is easily seen how these equations are solved. The solution is the following

$$p_n = \frac{1}{N} \sum_{0}^{N-1} \phi_1(m\delta)\omega^{-mn}, \qquad n = 0(1)N - 1.$$

If N is large the numerical work involved when you try to solve the equations in order to get the values of the coefficients p_n becomes, however, very large and it is here where the FFT algorithm comes in. With the aid of FFT these equations are solved rapidly.

Having arrived at the p-values we now define a periodic continuation

$$p_{n+N} = p_n.$$

We are now ready to present the approximate expression F for the distribution function corresponding to ϕ. It is found that this expression is equal to

$$\sum_{N/2}^{k-1} p_n + 0.5\ p_k = F(kd - Nd)$$

where k > N/2 and where we have adopted the principle that the value of the distribution function at a step is taken at half-way up the said step.

In order to examplify the use of the above formulas we take the following distribution function

$$p \in (x + 2) + (1 - p)(1 - e^{-x/a})$$

with

$$p = (\sqrt{41} - 5)/8$$

$$a = 2p/(1 = p).$$

The coefficients are chosen in such a way that the mean value is zero and the standard deviation 1. The characteristic function ϕ is now derived and we choose N = 1024 and d = 0.02. You will find below values calculated according to the above formulas

x	*Approximate value of distribution function*	*True value of distribution function*
-10.16	0.000000	0
-10.00	0.000000	0
- 9.84	0.000000	0
- 9.68	0.000000	0
- 9.52	0.000000	0
- 2.00	0.087695	0.087695
- 1.92	0.175391	0.175390
- 1.84	0.175391	0.175390
- 1.76	0.175391	0.175390
- 1.68	0.175391	0.175390
- 0.16	0.175404	0.175390
- 0.08	0.175447	0.175390
0.00	0.182680	0.175390
0.08	0.316690	0.316759
0.16	0.433801	0.433892
1.04	0.928454	0.928471
1.28	0.959301	0.959313
1.52	0.976847	0.976856
1.76	0.986827	0.986835
2.00	0.992504	0.992512

I hope that this example illustrates my point that the number of terms made possible by the FFT gives a very satisfactory result. In order to get still better approximation we have to take care of the aliason effect and the truncation effect, but this problem will not be dealt with here.

SOME PRACTICAL CONSIDERATIONS IN CONNECTION WITH THE CALCULATION OF STOP-LOSS PREMIUMS

Hans U. Gerber
Donald A. Jones

Department of Mathematics
The University of Michigan
Ann Arbor, Michigan

1. WHY THE EXPONENTIAL PRINCIPLE

A premium calculation principle is a rule that assigns a premium, say P, to any risk, say S. Mathematically, a risk is a random variable, given by its (supposedly sufficiently regular) distribution. The following four examples illustrate this concept.

a) *The net premium principle:*

$$P = E[S].$$

b) *The exponential principle:*

$$P = \frac{1}{a} \ln E[e^{a \cdot S}], \quad a > 0.$$

c) *The variance principle:*

$$P = E[S] + b \cdot Var[S], \quad b > 0.$$

d) *The standard deviation principle:*

$$P = E[S] + c \cdot \sqrt{Var[S]}, \quad c > 0.$$

ISBN 0-12-394680-8

In example a) there is no loading, in examples c), d) the loading is proportional to the variance or standard deviation, respectively. The exponential principle involves the evaluation of the moment generating function of S at the argument a.

The following two properties are highly desirable for a principle of premium calculation:

(P_1) $P \geq E[S]$

(P_2) $P \leq Max[S]$

for any risk S, where Max[S] denotes the right hand endpoint of the range of S. The first property means that the expected gain, P - E[S], is nonnegative for any risk S. The second property guarantees that the premium for any risk is not unreasonably high: If P > Max[S], the premium would exceed the maximal possible benefit, and nobody in the world would buy such a policy.

Obviously, the principles a), c), and d) above satisfy property (P_1). But so does the exponential principle: Jensen's inequality tells us that

$$E[e^{a \cdot S}] \geq e^{a \cdot E[S]}. \tag{1}$$

Now we take logarithms, divide by a, and recognize that (P_1) is satisfied for the exponential principle.

Principles a) and b) satisfy (P_2). The latter assertion follows from the inequality

$$E[e^{a \cdot S}] \leq e^{a \cdot Max[S]}. \tag{2}$$

Unfortunately, neither the variance principle nor the standard deviation principle satisfies property (P_2), as is seen from the following examples.

Example 1. Let S = 0 or Z, each with probability 1/2 (Z > 0). Thus Max[S] - Z. If the variance principle is applied, we find that

$$P = Z/2 + b \cdot Z^2/4. \tag{3}$$

But this means that $P > Max[S]$, whenever $Z > 2/b$.

Example 2. Let $S = 1$ (with probability p) or $S = 0$ (with probability $q = 1 - p$). Thus $Max\ S = 1$. If we apply the standard deviation principle, we get

$$P = p + c \cdot \sqrt{p(1-p)}. \tag{4}$$

By examining P as a function of p, we find that $P > Max[S]$, whenever $(1+c^2)^{-1} < p < 1$. Worse than that, if p is close to 1, P is a decreasing function of p!

The net premium principle and the exponential principle are *additive* as well as *iterative*; these concepts are explained in [1], p. 87 and p. 91. Under a mild continuity condition, these two principles can be characterized by these properties, see [3]. Also, the exponential principle fits into the framework of the collective theory of risk: The parameter "a" plays essentially the role of an adjustment coefficient.

For these reasons we shall adopt the exponential principle. Of course net premiums are always of interest; in fact they may be obtained as a limiting case $(a \to 0)$ from exponential premiums.

2. MEREU'S FORMULA, OR WHY THE TAIL IS NOT A PROBLEM

In the following we consider a stop-loss covarage (retention limit α) for aggregate claims X. Let F(x) denote the c d f of X. We assume that $F(x) = 0$ for $x < 0$ (no negative claims). The "risk" in question is now

$$S = (X-\alpha)_+ \begin{cases} 0 & \text{if } X \leq \alpha \\ X-\alpha & \text{if } X > \alpha \end{cases} \tag{5}$$

We denote the stop-loss premiums (based on principles a) and b) above) by $P(F,\alpha)$, $P(F,\alpha,a)$ respectively. Thus

$$P(F,\alpha) = E[(X-\alpha)+] = \int_{\alpha}^{\infty} (x-\alpha)\ dF(x) \tag{6}$$

and

$$P(F,\alpha,a) = \frac{1}{a}\ \ell n\ E[e^{a(x-\alpha)+}]$$

$$= \frac{1}{a}\ \ell n \left\{F(\alpha) + \int_{\alpha}^{\infty} e^{a(x-\alpha)} dF(x)\right\} \tag{7}$$

These formulas entail integration over the "tail" of F(x), which may cause computational problems. Fortunately, these difficulties can be avoided.

Mereu's idea [5] was to take expectations in the identity

$$(X-\alpha)+ = (X-\alpha) + (\alpha-X)+ \tag{8}$$

to get

$$P(F,\alpha) = E[X] - \alpha + E[(\alpha-X)+]$$

$$= E[X] - \alpha + \int_{0}^{\alpha} (\alpha-x)\ dF(x). \tag{9}$$

Thus if E[X] is obtainable otherwise, this formula requires the knowledge of F(x) for $0 < x < \alpha$ only.

In analogy to Mereu's idea let us now consider the identity

$$e^{a(X-\alpha)+} = e^{a(X-\alpha)} + \{1-e^{a(X-\alpha)}\}\ I_{[X<\alpha]}. \tag{10}$$

Then

$$E[e^{a(X-\alpha)+}] = E[e^{a(X-\alpha)}] + E[\{1-e^{a(X-\alpha)}\}\cdot I_{[X<\alpha]}]$$

$$= e^{-\alpha a}\ E[e^{aX}] + \int_{0}^{\alpha} (1-e^{a(x-\alpha)}) dF(x). \tag{11}$$

Thus

$$\underline{P}(F,\alpha,a) = \frac{1}{a}\ \ell n \left\{e^{-\alpha a} E[e^{aX}] + \int_{0}^{\alpha} (1-e^{a(x-\alpha)}) dF(x)\right\}. \tag{12}$$

Assuming that the moment generating function of X is available otherwise, this formula require the knowledge of F(x) for $0 < x < \alpha$ only, as before. Therefore formulas (9) and (12) are preferable to the original formulas (6) and (7).

Remark. By differentiating formula (11) k-times, and setting a = 0, one obtains an expression for the k-th absolute moment of $(X-\alpha)+$ in terms of the first k moments of X and the values of F(x) for $0 < x < \alpha$. In the case k = 1, this brings us back to formula (9). For k = 2, it leads to Mereu's formula (21), see [5].

3. THE COMPOUND POISSON DISTRIBUTION

In the following we assume that the aggregate claims X have a compound Poisson distribution, say with Poisson parameter λ (= expected number of claims) and jump amount distribution H(x) (= c d f of the individual claim amounts). Thus

$$F(x) = e^{-\lambda} \sum_{k=0}^{\infty} \frac{\lambda^k}{k!} H^{*k}(x). \tag{13}$$

It is assumed that H(0) = 0, which implies that F(x) = 0 for x < 0. In a certain sense, the assumption of a compound Poisson distribution is conservative (see the last paragraph in [2]).

If $\mu = \int x \, dH(x)$ denotes the average claim size, and $\phi(t) = \int e^{tx} \, dH(x)$ the moment generating function of the claim amounts,

$$\begin{aligned} E[X] &= \lambda \cdot \mu \\ E[e^{aX}] &= e^{\lambda[\phi(a)-1]} . \end{aligned} \tag{14}$$

These expressions should be substituted in formulas (9) and (12). It remains to calculate F(x) for $0 < x < \alpha$.

The latter problem is greatly simplified if H(x) is an arithmetic distribution, say with a span d > 0. This means that all possible claim amounts are multiples of the span d. Let h_i denote the probability that a given claim amount equals $i \cdot d$

$(i = 1,2,\ldots)$, and let $f_i = P[X = i\cdot d]$. Since

$$h_i^{*k} = 0 \quad \text{for} \quad k > i, \tag{15}$$

we see from formula (13) that

$$f_i = e^{-\lambda} \cdot \sum_{k=0}^{i} \frac{\lambda^k}{k!} h_i^{*k}. \tag{16}$$

For the evaluation of formulas (9) and (12) we have to compute h_i^{*k} and f_i for $i = 0,1,\ldots, [\frac{\alpha}{d}]$. Thus the larger the span d, or the smaller the retention limit α, the fewer calculations have to be made.

Remark. It is possible to calculate f_i by an algorithm that is somewhat different from the approach that results from formula (16). Let N_j denote the number of claims of amount $j\cdot d$. Obviously,

$$X = d\cdot N_1 + 2d\cdot N_2 + 3d\cdot N_3 + \ldots \tag{17}$$

Also, it is well known that N_j has a Poisson distribution with parameter $\lambda\cdot h_j$, and that the random variables $N_1, N_2, \ldots$ are mutually independent. From this it follows that

$$f_i = (p^{(1)} * p^{(2)} * p^{(3)} * \ldots)_i \tag{18}$$

where $p_i^{(j)} = P[j\cdot N_j = i]$, or

$$p_i^{(j)} = \begin{cases} e^{-\lambda h_j} \cdot \dfrac{(\lambda h_j)^m}{m!} & \text{if } m = \dfrac{i}{j} \text{ is an integer} \\ 0 \quad \text{if } m = \dfrac{i}{j} \text{ is not an integer.} & \end{cases} \tag{19}$$

For the calculation of f_i $(i = 0,1,\ldots, r = [\frac{\alpha}{d}])$, one needs $f_i^{(j)}$ for $i = 0,1,\ldots,[\frac{r}{j}]$. Thus if $j > r$, we need only

$$p_o^{(j)} = e^{-\lambda h_j}. \tag{20}$$

Therefore, formula (18) can be rewritten as follows:

$$f_i = (p^{(1)} * \ldots * p^{(r)})_i \cdot e^{-\lambda} \sum_{r+1}^{\infty} h_j. \tag{21}$$

An algorithm similar to this has been developed in [4].

4. THE METHOD OF DISPERSAL

In the last section we saw that the exact calculation of a stop-loss premium is feasible if the claim amount distribution is arithmetic with a sufficiently large span. If F is an arbitrary compound Poisson distribution (non-arithmetic, or arithmetic with a small span), the exact calculation of the stop-loss premium may be an extensive procedure and may lead to considerable round-off errors. Instead, we suggest a procedure that is outlined in this and the following section. The idea is to replace the original distribution F by arithmetic compound Poisson distributions F^u, F , respectively, calculating the stop-loss premiums for these, thereby getting upper and lower bounds for P(F, ,a).

We pick a span $d > 0$. Then we construct a compound Poisson distribution F^u, given by its Poisson parameter λ^u and the jump amount distribution $H^u(x)$, as follows: we set $\lambda^u = \lambda$, and H^u is arithmetic with span d, such that

$$h_i^u = \int_{(i-1)d}^{(i+1)d} \left(1 - \left|\frac{x}{d} - 1\right|\right) dH(x) \tag{22}$$

for $i = 0, 1, 2, \ldots$. This simply means that the probability mass of H(x) between $i \cdot d$ and $(i + 1)d$ is dispersed to the end-points $i \cdot d$ and $(i + 1)d$, such that the conditional mean remains unchanged (for $i = 0, 1, 2, \cdots$). Consequently, the mean of F^u equals the mean of F.

Theorem 1. $P(F,\alpha,a) \leq P(F^u,\alpha,a)$ for all $a \geq 0$, α. Thus the method of dispersal leads (for each d) to an upper bound for the stop-loss premium. Theorem 1 can be proven by using (in this order)

The second part of Example 2 in [2], with $a = i \cdot d$, $b = (i + 1)d$, applied to the conditional distribution of $H(x)$, $a \leq x < b$.

Theorem 2 in [2].

Lemma 1 in [2] to show that $H(x) < H^u(x)$.

Lemma 3 and Lemma 1 in [2] to show that $F(x) < F^u(x)$.

Theorem 1 in [2] to complete the proof.

Here our concern is with the content of the theorem, but not with its proof.

5. THE METHOD OF TRUNCATION

As before, we first pick a span $d > 0$. Then we associate a compound Poisson distribution F^ℓ (Poisson parameter λ^ℓ, jump amount distribution H^ℓ arithmetic with span d) to the original distribution F (given by λ and H) as follows:

$$\lambda^\ell h_i^\ell = \lambda \int_{i \cdot d}^{(i+1)d} \frac{x}{i \cdot d} \, dH(x) \tag{23}$$

for $i = 1,2,\cdots$. The value of λ^ℓ is determined by the condition that $h_1^\ell + h_2^\ell + \cdots = 1$. The underlying idea is the following: if $i \cdot d \leq x < (i + 1)d$, a claim of amount x is replaced by one of amount $i \cdot d$. In order to keep expected values unchanged, we compensate by increasing the Poisson frequency proportionately. In this truncation process, the claims between 0 and d are ignored. Thus

$$\int_0^\infty x \, dF(x) - \int_0^\infty x \, dF^\ell(x) = \lambda \int_0^d x \, dH(x). \tag{24}$$

Truncation leads to a lower bound for the stop-loss premium:

Theorem 2. $P(F,\alpha,a) \geq P(F^\ell,\alpha,a)$ for all $a \geq 0,\alpha$.

The proof is similar to the one for Theorem 1, with the following modification: In the first step, part 1 of Example 2 in [2] is applied, with $a = 0$ and $b = (i + 1)d$, to the mixture of the degenerate distribution concentrated at 0 and the conditional distribution of $H(x)$, given $i \cdot d \leq x < (i + 1)d$, where the weights are chosen such that the mean of this mixture equals $i \cdot d$ $(i = 1, 2, \cdots)$.

6. AN ILLUSTRATIVE EXAMPLE

To illustrate the dispersal and truncation methods, let us consider the sample portfolio of 5 policies that is defined in Table 1. Thus for the compound Poisson distribution F we have $\lambda = 1.4$, and the claim amount distribution is concentrated at the arguments 1.7, 2.3, etc., with probabilities .2/1.4, .3/1.4, etc. Thus the original claim amount distribution is arithmetic with a span $d = .1$.

The methods of dispersal and truncation are best applied policy by policy. Table 2 shows this procedure for $d = 1$. The data on the bottom line are used to calculate $P(F^u,\alpha,a)$ and $P(F^\ell,\alpha,a)$ which are shown in Tables 5 and 6. For $d = 2$ the calculations are summarized in Table 3.

Table 1.

Policy	*Amount at risk*	*Mortality rate*
A	*$1.7*	*.2*
B	*$2.3*	*.3*
C	*$3.4*	*.3*
D	*$3.6*	*.4*
E	*$5.0*	*.2*
Total		*1.4*

Table 2.

	Contribution to									
Policy	$\lambda^u h_1^u$	$\lambda^u h_2^u$	$\lambda^u h_3^u$	$\lambda^u h_4^u$	$\lambda^u h_5^u$	$\lambda^\ell h_1^\ell$	$\lambda^\ell h_2^\ell$	$\lambda^\ell h_3^\ell$	$\lambda^\ell h_4^\ell$	$\lambda^\ell h_5^\ell$
A	.06	.14				.34				
B		.21	.09				.345			
C			.18	.12				.34		
D			.16	.24				.48		
E					.20					.20
Total	.06	.35	.43	.36	.20	.34	.345	.82	---	.20

Table 3.

	Contribution to					
Policy	$\lambda^u h_0^u$	$\lambda^u h_1^u$	$\lambda^u h_2^u$	$\lambda^u h_3^u$	$\lambda^\ell h_1^\ell$	$\lambda^\ell h_2^\ell$
A	.03	.17				
B		.255	.045		.345	
C		.09	.21		.51	
D		.08	.32		.72	
E			.10	.10		.25
Total	.03	.595	.675	.10	1.575	.25

Table 4.

Amount at risk	$\lambda^u h_i^u$		$\lambda^u h_i^u$	
	$d = 1$	$d = 2$	$d = 1$	$d = 2$
\$ 0		.03		*dropped*
\$ 1	.06		.34	
\$ 2	.35	.35+.03+.215	.345	$.345+\frac{3}{2}(.82)$
\$ 3	.43		.82	
\$ 4	.36	.36+.215+.10		$\frac{5}{4}(.20)$
\$ 5	.20		.20	
\$ 6		.10		

However, in the case of a large number of policies, it is more economical to obtain the bottom line of Table 3 directly from the bottom line of Table 2. The calculations are displayed in Table 4.

In general, this shortcut works if the second span is a multiple of the first span. From this observation, and from Theorem 1 and 2, respectively, it follows that if the original span is replaced by a multiplie of it, the upper bound is increased, while the lower bound is decreased. In practice one uses this in the opposite direction: if for a given span the upper bound differs from the lower bound by too much, one may want to replace this span by one nth of it.

In Tables 5 and 6 the numerical results are shown for one sample portfolio for a span d = 1. The first column shows the "amounts". These should be interpreted as the arguments of the frequency function of the aggregate claims (column two) and their cumulative distribution function (column three) or the one hand, and as retention limits for the net stop-loss premiums (column four) and the exponential stop-loss premiums (column five) on the other hand. The latter are based on a = .1. Tables 7 and 8 display the results for $d = 2$. Of course our sample portfolio is small enough so that the exact distribution of aggregate claims and the exact stop-loss premiums can be calculated. Table 9 shows these values calculated as lower bound for $d = 0.1$ which are exact.

For the convenience of the reader, Tables 10 and 11 provide comparisons of the bounds for the premiums from Tables 5 - 8 with the exact values from Table 9. Tables 10 and 11 show the relative errors of the bounds increasing as the retention level increases - but remember that all are going to zero.

TABLE 5. Upper Bounds (D = 1.00) for Net and Exponential (A = 0.10) Stop Loss Premiums

AMOUNT	FREQUENCY	CUMULATIVE	NET PREMIUM	EXP PREMIUM
0.0	0.246597	0.246597	4.490000	5.410417
1.00	0.014796	0.261393	3.736597	4.560266
2.00	0.086753	0.348146	2.997990	3.733002
3.00	0.111224	0.459370	2.346135	2.981955
4.00	0.110397	0.569766	1.805505	2.334229
5.00	0.092859	0.662625	1.375271	1.797797
6.00	0.061008	0.723633	1.037897	1.363697
7.00	0.065427	0.789060	0.761530	1.006613
8.00	0.054577	0.843637	0.550590	0.730186
9.00	0.041321	0.884958	0.394228	0.522717
10.00	0.030579	0.915537	0.279186	0.369178
11.00	0.023308	0.938845	0.194723	0.256592
12.00	0.018344	0.957189	0.133568	0.175434
13.00	0.013149	0.970338	0.090757	0.118692
14.00	0.009218	0.979556	0.061096	0.079494
15.00	0.006504	0.986061	0.040652	0.052622
16.00	0.004596	0.990656	0.026713	0.034416
17.00	0.003176	0.993833	0.017369	0.022278
18.00	0.002123	0.995956	0.011202	0.014301
19.00	0.001414	0.997370	0.007158	0.009097
20.00	0.000940	0.998309	0.004528	0.005731
21.00	0.000617	0.998926	0.002837	0.003577
22.00	0.000398	0.999324	0.001764	0.002216
23.00	0.000253	0.999578	0.001088	0.001362
24.00	0.000161	0.999738	0.000666	0.000830
25.00	0.000101	0.999839	0.000404	0.000502
26.00	0.000063	0.999902	0.000243	0.000301
27.00	0.000039	0.999941	0.000145	0.000180
28.00	0.000024	0.999965	0.000086	0.000106
29.00	0.000014	0.999979	0.000051	0.000063
30.00	0.000009	0.999988	0.000030	0.000037
31.00	0.000005	0.999993	0.000017	0.000021
32.00	0.000003	0.999996	0.000010	0.000012
33.00	0.000002	0.999998	0.000006	0.000007
34.00	0.000001	0.999999	0.000003	0.000004
35.00	0.000001	0.999999	0.000002	0.000002
36.00	0.000000	1.000000	0.000001	0.000001

TABLE 6. *Lower Bounds (D = 1.00) for Net and Exponential (A = 0.10) Stop Loss Premiums*

AMOUNT	FREQUENCY	CUMULATIVE	NET PREMIUM	EXP PREMIUM
0.0	0.181772	0.181772	4.490000	5.287705
1.00	0.061803	0.243575	3.671772	4.399739
2.00	0.073218	0.316793	2.915347	3.563379
3.00	0.171566	0.488359	2.232140	2.794000
4.00	0.065222	0.553581	1.720499	2.175059
5.00	0.100489	0.654070	1.274080	1.632818
6.00	0.093837	0.747907	0.928149	1.200648
7.00	0.047844	0.795751	0.676056	0.874981
8.00	0.062473	0.858224	0.471807	0.613788
9.00	0.038924	0.897147	0.330031	0.428544
10.00	0.027452	0.924600	0.227178	0.293951
11.00	0.025792	0.950392	0.151778	0.196235
12.00	0.014276	0.964667	0.102170	0.131241
13.00	0.011743	0.976410	0.066837	0.085544
14.00	0.008301	0.984711	0.043247	0.055135
15.00	0.004900	0.989611	0.027959	0.035414
16.00	0.003880	0.993491	0.017570	0.022189
17.00	0.002317	0.995808	0.011060	0.013896
18.00	0.001514	0.997323	0.006868	0.008589
19.00	0.001050	0.998373	0.004191	0.005226
20.00	0.000600	0.998973	0.002564	0.003181
21.00	0.000406	0.999379	0.001537	0.001901
22.00	0.000248	0.999627	0.000916	0.001130
23.00	0.000146	0.999773	0.000544	0.000668
24.00	0.000095	0.999868	0.000317	0.000389
25.00	0.000054	0.999922	0.000185	0.000226
26.00	0.000033	0.999954	0.000107	0.000130
27.00	0.000020	0.999974	0.000061	0.000074
28.00	0.000011	0.999985	0.000035	0.000042
29.00	0.000007	0.999991	0.000019	0.000024
30.00	0.000004	0.999995	0.000011	0.000013
31.00	0.000002	0.999997	0.000006	0.000007
32.00	0.000001	0.999998	0.000003	0.000004
33.00	0.000001	0.999999	0.000002	0.000002
34.00	0.000000	1.000000	0.000001	0.000001
35.00	0.000000	1.000000	0.000001	0.000001
36.00	0.000000	1.000000	0.000000	0.000000

TABLE 7. Upper Bounds (D = 2.00) for Net and Exponential (A = 0.10) Stop Loss Premiums

AMOUNT	FREQUENCY	CUMULATIVE	NET PREMIUM	EXP PREMIUM
0.0	0.254107	0.254107	4.490000	5.459282
1.00	0.0	0.254107	3.744107	4.612913
2.00	0.151194	0.405301	2.998214	3.780000
3.00	0.0	0.405301	2.403515	3.067901
4.00	0.216502	0.621803	1.808815	2.376726
5.00	0.0	0.621803	1.430618	1.879491
6.00	0.136387	0.758190	1.052421	1.407223
7.00	0.0	0.758190	0.810611	1.077005
8.00	0.104697	0.862887	0.568802	0.768514
9.00	0.0	0.862887	0.431689	0.575439
10.00	0.062274	0.925161	0.294576	0.397467
11.00	0.0	0.925161	0.219736	0.291380
12.00	0.036552	0.961712	0.144897	0.194409
13.00	0.0	0.961712	0.106609	0.140207
14.00	0.019604	0.981316	0.068322	0.090908
15.00	0.0	0.981316	0.049638	0.064642
16.00	0.009961	0.991277	0.030954	0.040817
17.00	0.0	0.991277	0.022232	0.028666
18.00	0.004818	0.996095	0.013509	0.017659
19.00	0.0	0.996095	0.009604	0.012269
20.00	0.002220	0.998315	0.005699	0.007390
21.00	0.0	0.998315	0.004014	0.005084
22.00	0.000983	0.999298	0.002328	0.002997
23.00	0.0	0.999298	0.001626	0.002043
24.00	0.000419	0.999716	0.000923	0.001181
25.00	0.0	0.999716	0.000640	0.000798
26.00	0.000172	0.999889	0.000356	0.000452
27.00	0.0	0.999889	0.000245	0.000304
28.00	0.000069	0.999958	0.000134	0.000169
29.00	0.0	0.999958	0.000091	0.000113
30.00	0.00027	0.999984	0.000049	0.000062
31.00	0.0	0.999984	0.000033	0.000041
32.00	0.000010	0.999994	0.000018	0.000022
33.00	0.0	0.999994	0.000012	0.000014
34.00	0.000004	0.999998	0.000006	0.000008
35.00	0.0	0.999998	0.000004	0.000005
36.00	0.000001	0.999999	0.000002	0.000003

TABLE 8. *Lower Bounds (D = 2.00) for Net and Exponential (A = 0.10) Stop Loss Premiums*

AMOUNT	FREQUENCY	CUMULATIVE	NET PREMIUM	EXP PREMIUM
0.0	0.161218	0.161218	4.150000	4.716655
1.00	0.0	0.161218	3.311218	3.821895
2.00	0.253918	0.415135	2.472435	2.936929
3.00	0.0	0.415135	1.887571	2.257233
4.00	0.240265	0.655400	1.302706	1.599683
5.00	0.0	0.655400	0.958106	1.170472
6.00	0.168459	0.823859	0.613506	0.765562
7.00	0.0	0.823859	0.437365	0.537582
8.00	0.096364	0.920222	0.261224	0.326720
9.00	0.0	0.920222	0.181446	0.222107
10.00	0.047200	0.967423	0.101668	0.126497
11.00	0.0	0.967423	0.069091	0.083833
12.00	0.010420	0.987843	0.036514	0.045071
13.00	0.0	0.987843	0.024357	0.029262
14.00	0.007966	0.995809	0.012200	0.014936
15.00	0.0	0.995809	0.008009	0.009533
16.00	0.002845	0.998654	0.003819	0.004640
17.00	0.0	0.998654	0.002472	0.002918
18.00	0.000940	0.999594	0.001126	0.001360
19.00	0.0	0.999594	0.000720	0.000844
20.00	0.000290	0.999884	0.000315	0.000378
21.00	0.0	0.999884	0.000199	0.000232
22.00	0.000084	0.999969	0.000084	0.000100
23.00	0.0	0.999969	0.000052	0.000061
24.00	0.000023	0.999992	0.000021	0.000025
25.00	0.0	0.999992	0.000013	0.000015
26.00	0.000006	0.999998	0.000005	0.000006
27.00	0.0	0.999998	0.000003	0.000004
28.00	0.000002	1.000000	0.000001	0.000001
29.00	0.0	1.000000	0.000001	0.000001
30.00	0.000000	1.000000	0.000000	0.000000
31.00	0.0	1.000000	0.000000	0.000000
32.00	0.000000	1.000000	0.000000	0.000000
33.00	0.0	1.000000	0.000000	0.000000
34.00	0.000000	1.000000	0.000000	0.000000
35.00	0.0	1.000000	0.000000	0.000000
36.00	0.000000	1.000000	0.000000	0.000000

TABLE 9. Lower Bounds (D = 0.10) for Net and Exponential (A = 0.10) Stop Loss Premiums

AMOUNT	FREQUENCY	CUMULATIVE	NET PREMIUM	EXP PREMIUM
0.0	0.246597	0.246597	4.490000	5.392013
0.10	0.0	0.246597	4.414660	5.306456
0.20	0.0	0.246597	4.339319	5.221024
0.30	0.0	0.246597	4.263979	5.135716
0.40	0.0	0.246597	4.188639	5.050535
0.50	0.0	0.246597	4.113298	4.965480
0.60	0.0	0.246597	4.037958	4.880552
0.70	0.0	0.246597	3.962618	4.795753
0.80	0.0	0.246597	3.887278	4.711083
0.90	0.0	0.246597	3.811937	4.626544
1.00	0.0	0.246597	3.736597	4.542136
1.10	0.0	0.246597	3.661257	4.457860
1.20	0.0	0.246597	3.585916	4.373716
1.30	0.0	0.246597	3.510576	4.289707
1.40	0.0	0.246597	3.435236	4.205832
1.50	0.0	0.246597	3.359895	4.122093
1.60	0.0	0.246597	3.284555	4.038491
1.70	0.049319	0.295916	3.209215	3.955027
1.80	0.0	0.295916	3.138806	3.875032
1.90	0.0	0.295916	3.068398	3.795197
2.00	0.0	0.295916	2.997990	3.715525
2.10	0.0	0.295916	2.927581	3.636014
2.20	0.0	0.295916	2.857173	3.556667
2.30	0.073979	0.369895	2.786765	3.477485
2.40	0.0	0.369895	2.723754	3.403706
2.50	0.0	0.369895	2.660744	3.330122
2.60	0.0	0.369895	2.597733	3.256732
2.70	0.0	0.369895	2.534723	3.183538
2.80	0.0	0.369895	2.471712	3.110541
2.90	0.0	0.369895	2.408702	3.037741
3.00	0.0	0.369895	2.345691	2.965139
3.10	0.0	0.369895	2.282681	2.892737
3.20	0.0	0.369895	2.219671	2.820536
3.30	0.0	0.369895	2.156660	2.748535
3.40	0.078911	0.448806	2.093650	2.676737
3.50	0.0	0.448806	2.038530	2.611190
3.60	0.098639	0.547445	1.983411	2.545870
3.70	0.0	0.547445	1.938155	2.488432
3.80	0.0	0.547445	1.892900	2.431239
3.90	0.0	0.547445	1.8477645	2.374291
4.00	0.014796	0.562241	1.802389	2.317588
4.10	0.0	0.562241	1.758613	2.262305
4.20	0.0	0.562241	1.714837	2.207270
4.30	0.0	0.562241	1.671061	2.152481

TABLE 9 (cont.)

AMOUNT	FREQUENCY	CUMULATIVE	NET PREMIUM	EXP PREMIUM
4.40	0.0	0.562241	1.627285	2.097941
4.50	0.0	0.562241	1.583510	2.043649
4.60	0.011097	0.573338	1.539734	1.989605
4.70	0.0	0.573338	1.497067	1.936719
4.80	0.0	0.573338	1.454401	1.884083
4.90	0.0	0.573338	1.411735	1.831696
5.00	0.049319	0.622657	1.369069	1.779558
5.10	0.015125	0.637782	1.331335	1.731798
5.20	0.0	0.637782	1.295113	1.685559
5.30	0.019728	0.657510	1.258891	1.639569
5.40	0.0	0.657510	1.224642	1.595500
5.50	0.0	0.657510	1.190393	1.551677
5.60	0.0	0.657510	1.156144	1.508101
5.70	0.023673	0.681183	1.121895	1.464770
5.80	0.0	0.681183	1.090013	1.423728
5.90	0.029592	0.710775	1.058131	1.382928
6.00	0.0	0.710775	1.029209	1.344943
6.10	0.0	0.710775	1.000286	1.307194
6.20	0.0	0.710775	0.971364	1.269679
6.30	0.002219	0.712994	0.942441	1.232398
6.40	0.0	0.712994	0.913741	1.195547
6.50	0.0	0.712994	0.885040	1.158928
6.60	0.0	0.712994	0.856340	1.122541
6.70	0.009864	0.722858	0.827639	1.086385
6.80	0.012593	0.735451	0.799925	1.051343
6.90	0.001110	0.736560	0.773470	1.017661
7.00	0.031564	0.768125	0.747126	0.984301
7.10	0.0	0.768125	0.723938	0.954019
7.20	0.019728	0.787853	0.700751	0.923947
7.30	0.014796	0.802648	0.679536	0.895880
7.40	0.004537	0.807186	0.659801	0.869364
7.50	0.0	0.807186	0.640520	0.843458
7.60	0.005918	0.813104	0.621238	0.817743
7.70	0.0	0.813104	0.602549	0.792763
7.80	0.0	0.813104	0.583859	0.767969
7.90	0.0	0.813104	0.565169	0.743362
8.00	0.003551	0.816655	0.546480	0.718940
8.10	0.0	0.816655	0.528145	0.695031
8.20	0.004439	0.821094	0.509811	0.671304
8.30	0.0	0.821094	0.491920	0.648171
8.40	0.015782	0.836876	0.474030	0.625216
8.50	0.002319	0.839195	0.457717	0.603915
8.60	0.019950	0.859144	0.441637	0.582999
8.70	0.006050	0.865194	0.427551	0.564125
8.80	0.0	0.865194	0.414071	0.545973
8.90	0.003946	0.869140	0.400590	0.527969

TABLE 9 (cont.)

AMOUNT	FREQUENCY	CUMULATIVE	NET PREMIUM	EXP PREMIUM
9.00	0.002959	0.872099	0.387504	0.510486
9.10	0.003778	0.875877	0.374714	0.493426
9.20	0.000083	0.875960	0.362302	0.476866
9.30	0.009469	0.885429	0.349898	0.460452
9.40	0.0	0.885429	0.338441	0.445075
9.50	0.005918	0.891348	0.326983	0.429829
9.60	0.002219	0.893567	0.316118	0.415275
9.70	0.000681	0.894248	0.305475	0.401058
9.80	0.0	0.894248	0.294900	0.387028
9.90	0.000888	0.89513	0.284325	0.373118
10.00	0.004932	0.900067	0.273838	0.359412
10.10	0.003025	0.903092	0.263845	0.346299
10.20	0.001337	0.904429	0.254154	0.333590
10.30	0.004301	0.908730	0.244597	0.321120
10.40	0.005037	0.913767	0.235470	0.309174
10.50	0.000444	0.914211	0.226847	0.297819
10.60	0.006313	0.920524	0.218268	0.286607
10.70	0.004735	0.925258	0.210320	0.276106
10.80	0.003326	0.928584	0.202846	0.266157
10.90	0.005935	0.934519	0.195704	0.256619
11.00	0.001815	0.936334	0.189156	0.247744
11.10	0.0	0.936334	0.182790	0.239126
11.20	0.001184	0.937518	0.176423	0.230586
11.30	0.000444	0.937962	0.170175	0.222239
11.40	0.000567	0.938528	0.163971	0.214011
11.50	0.000005	0.938533	0.157824	0.205914
11.60	0.001420	0.939954	0.151677	0.197892
11.70	0.000986	0.940940	0.145672	0.190081
11.80	0.003406	0.944346	0.139766	0.182439
11.90	0.000459	0.944805	0.134201	0.175200
12.00	0.006381	0.951186	0.128682	0.168073
12.10	0.000927	0.952114	0.123800	0.161636
12.20	0.004034	0.956148	0.119012	0.255350
12.30	0.002690	0.958838	0.114626	0.149519
12.40	0.000907	0.959745	0.110510	0.144006
12.50	0.000927	0.960672	0.106485	0.138634
12.60	0.001210	0.961882	0.102552	0.133403
12.70	0.001511	0.963394	0.098740	0.128341
12.80	0.000033	0.963427	0.095079	0.123475
12.90	0.001894	0.965321	0.091422	0.118658
13.00	0.000710	0.966031	0.087954	0.114073
13.10	0.000893	0.966924	0.084557	0.109602
13.20	0.000889	0.967813	0.081250	0.105261
13.30	0.000272	0.968085	0.078031	0.101049
13.40	0.001578	0.969664	0.074840	0.096904
13.50	0.000641	0.970305	0.071806	0.092955

TABLE 9. (cont.)

AMOUNT	FREQUENCY	CUMULATIVE	NET PREMIUM	EXP PREMIUM
13.60	0.002123	0.972428	0.068836	0.089106
13.70	0.001267	0.973695	0.066079	0.085504
13.80	0.000535	0.974230	0.063449	0.082061
13.90	0.000931	0.975161	0.060872	0.078704
14.00	0.001303	0.976464	0.058388	0.075471
14.10	0.000844	0.977309	0.056034	0.072398
14.20	0.000929	0.978238	0.053765	0.069438
14.30	0.001899	0.980137	0.051589	0.066599
14.40	0.000541	0.980678	0.049602	0.063975
14.50	0.001190	0.981869	0.047670	0.061430
14.60	0.000585	0.982453	0.045857	0.059027
14.70	0.000136	0.982590	0.044103	0.056706
14.80	0.000218	0.982808	0.042361	0.054420
14.90	0.000179	0.982987	0.040642	0.052178
15.00	0.000555	0.983542	0.038941	0.049976
15.10	0.000304	0.983847	0.037295	0.047851
15.20	0.000551	0.984398	0.035680	0.045776
15.30	0.000484	0.984882	0.034120	0.043776
15.40	0.001136	0.986018	0.032608	0.041843
15.50	0.000184	0.986202	0.031210	0.040042
15.60	0.001290	0.987491	0.029830	0.038277
15.70	0.000659	0.988150	0.028579	0.036657
15.80	0.000683	0.988833	0.027394	0.035118
15.90	0.000788	0.989622	0.026277	0.033662
16.00	0.000367	0.989989	0.025239	0.032298
16.10	0.000213	0.990202	0.024238	0.030984
16.20	0.000247	0.990449	0.023258	0.029704
16.30	0.000347	0.990796	0.022303	0.028462
16.40	0.000120	0.990916	0.021383	0.027266
16.50	0.000264	0.991180	0.020475	0.026093
16.60	0.000284	0.991464	0.019593	0.024959
16.70	0.000186	0.991651	0.018739	0.023863
16.80	0.000480	0.992081	0.017904	0.022797
16.90	0.000124	0.992205	0.017112	0.021785
17.00	0.000652	0.992856	0.016333	0.020794
17.10	0.000215	0.993072	0.015618	0.019878
17.20	0.000455	0.993526	0.014925	0.018992
17.30	0.000363	0.993890	0.014278	0.018161
17.40	0.000198	0.994087	0.013667	0.017373
17.50	0.000214	0.994301	0.013076	0.016613
17.60	0.000263	0.994565	0.012506	0.015882
17.70	0.000315	0.994880	0.011962	0.015184
17.80	0.000119	0.994999	0.011450	0.014524
17.90	0.000381	0.995380	0.010950	0.013883
18.00	0.000148	0.995527	0.010488	0.013286
18.10	0.00018	0.995707	0.010041	0.012709

TABLE 9 (cont.)

AMOUNT	FREQUENCY	CUMULATIVE	NET PREMIUM	EXP PREMIUM
18.20	0.000142	0.995850	0.009612	0.012156
18.30	0.000055	0.995904	0.009197	0.011623
18.40	0.000145	0.996049	0.008787	0.011100
18.50	0.000033	0.996132	0.008392	0.010597
18.60	0.000203	0.996334	0.008005	0.010107
18.70	0.000134	0.996468	0.007639	0.009642
18.80	0.000146	0.996614	0.007285	0.009195
18.90	0.000115	0.996729	0.006947	0.008767
19.00	0.000237	0.996966	0.006620	0.008355
19.10	0.000112	0.997078	0.006316	0.007970
19.20	0.000190	0.997268	0.006024	0.007600
19.30	0.000217	0.997485	0.005751	0.007253
19.40	0.000111	0.997596	0.005499	0.006931
19.50	0.000149	0.997745	0.005259	0.006623
19.60	0.000089	0.997834	0.005033	0.006333
19.70	0.000056	0.997884	0.004817	0.006054
19.80	0.000046	0.997929	0.004605	0.005784
19.90	0.000059	0.997989	0.004398	0.005520
20.00	0.000063	0.998051	0.004197	0.005265
20.10	0.000050	0.998101	0.004002	0.005019
20.20	0.000084	0.998186	0.003812	0.004780
20.30	0.000052	0.998237	0.003631	0.004552
20.40	0.000127	0.998364	0.003454	0.004332
20.50	0.000036	0.998399	0.003291	0.004126
20.60	0.000134	0.998534	0.003131	0.003926
20.70	0.000073	0.998607	0.002984	0.003741
20.80	0.000079	0.998686	0.002845	0.003565
20.90	0.000083	0.998769	0.002713	0.003399
21.00	0.000052	0.998820	0.002590	0.003242
21.10	0.000046	0.998867	0.002472	0.003093
21.20	0.000041	0.998908	0.002359	0.002949
21.30	0.000065	0.998973	0.002250	0.002811
21.40	0.000025	0.998998	0.002147	0.002681
21.50	0.000053	0.999051	0.002047	0.002555
21.60	0.000038	0.999089	0.001952	0.002435
21.70	0.000028	0.999117	0.001861	0.002320
21.80	0.000041	0.999158	0.001773	0.002209
21.90	0.000017	0.999175	0.001689	0.002104
22.00	0.000051	0.999226	0.001606	0.002001
22.10	0.000025	0.999251	0.001529	0.001904
22.20	0.000043	0.999294	0.001454	0.001810
22.30	0.000034	0.999328	0.001383	0.001722
22.40	0.000032	0.999359	0.001316	0.001638
22.50	0.000026	0.999385	0.001252	0.001558
22.60	0.000038	0.999423	0.001190	0.001481
22.70	0.000035	0.999459	0.001133	0.001409

TABLE 9 (cont.)

AMOUNT	FREQUENCY	CUMULATIVE	NET PREMIUM	EXP PREMIUM
22.80	0.000024	0.999483	0.001078	0.001341
22.90	0.000042	0.999524	0.001027	0.001276
23.00	0.000020	0.999545	0.000979	0.001216
23.10	0.000022	0.999567	0.000934	0.001159
23.20	0.000017	0.999583	0.000890	0.001104
23.30	0.000010	0.999594	0.000849	0.001052
23.40	0.000014	0.999607	0.000808	0.001001
23.50	0.000011	0.999618	0.000769	0.000952
23.60	0.000018	0.999636	0.000730	0.000904
23.70	0.000012	0.999649	0.000694	0.000859
23.80	0.000019	0.999667	0.000659	0.000816
23.90	0.000011	0.999678	0.000626	0.000775
24.00	0.000024	0.999703	0.000594	0.000735
24.10	0.000012	0.999714	0.000564	0.000698
24.20	0.000026	0.999734	0.000535	0.000663
24.30	0.000019	0.999753	0.000509	0.000630
24.40	0.000013	0.999766	0.000484	0.000599
24.50	0.000014	0.999780	0.000461	0.000570
24.60	0.000010	0.999790	0.000439	0.000542
24.70	0.000009	0.999799	0.000418	0.000516
24.80	0.000006	0.999806	0.000397	0.000491
24.90	0.000010	0.999814	0.000378	0.000466
25.00	0.000007	0.999821	0.000359	0.000443
25.10	0.000007	0.999828	0.000342	0.000421
25.20	0.000008	0.999837	0.000324	0.000400
25.30	0.000005	0.999842	0.000308	0.000379
25.40	0.000010	0.999852	0.000292	0.000360
25.50	0.000004	0.999856	0.000277	0.000342
25.60	0.000010	0.999866	0.000263	0.000324
25.70	0.000006	0.999872	0.000250	0.000307
25.80	0.000007	0.999880	0.000237	0.000291
25.90	0.000007	0.999886	0.000225	0.000277
26.00	0.000006	0.999892	0.000213	0.000263
26.10	0.000005	0.999897	0.000203	0.000249
26.20	0.000005	0.999903	0.000192	0.000237
26.30	0.000007	0.999909	0.000183	0.000224
26.40	0.000003	0.999913	0.000173	0.000213
26.50	0.000006	0.999918	0.000165	0.000202
26.60	0.000004	0.999922	0.000157	0.000192
26.70	0.000003	0.999925	0.000149	0.000183
26.80	0.000003	0.999929	0.000141	0.000173
26.90	0.000002	0.999931	0.000134	0.000165
27.00	0.000004	0.999934	0.000127	0.000156
27.10	0.000002	0.999936	0.000121	0.000148
27.20	0.000004	0.999940	0.000114	0.000140
27.30	0.000003	0.999943	0.000108	0.000133

TABLE 9 (cont.)

AMOUNT	FREQUENCY	CUMULATIVE	NET PREMIUM	EXP PREMIUM
27.40	0.000003	0.999946	0.000103	0.000126
27.50	0.000002	0.999948	0.000097	0.000119
27.60	0.000004	0.999952	0.000092	0.000113
27.70	0.000003	0.999955	0.000087	0.000107
27.80	0.000003	0.999957	0.000083	0.000101
27.90	0.000003	0.999960	0.000078	0.000096
28.00	0.000002	0.999962	0.000074	0.000091
28.10	0.000002	0.999964	0.000071	0.000086
28.20	0.000002	0.999966	0.000067	0.000082
28.30	0.000001	0.999967	0.000064	0.000078
28.40	0.000001	0.999969	0.000060	0.000074
28.50	0.000001	0.999970	0.000057	0.000070
28.60	0.000001	0.999971	0.000054	0.000066
28.70	0.000001	0.999972	0.000051	0.000063
28.80	0.000002	0.999974	0.000049	0.000059
28.90	0.000001	0.999975	0.000046	0.000056
29.00	0.000002	0.999977	0.000044	0.000053
29.10	0.000001	0.999978	0.000041	0.000050
29.20	0.000002	0.999979	0.000039	0.000048
29.30	0.000001	0.999980	0.000037	0.000045
29.40	0.000001	0.999982	0.000035	0.000043
29.50	0.000001	0.999983	0.000033	0.000040
29.60	0.000001	0.999984	0.000031	0.000038
29.70	0.000001	0.999984	0.000030	0.000036
29.80	0.000001	0.999985	0.000028	0.000034
29.90	0.000001	0.999986	0.000027	0.000033
30.00	0.000001	0.999987	0.000025	0.000031
30.10	0.000001	0.999987	0.000024	0.000029
30.20	0.000001	0.999988	0.000023	0.000028
30.30	0.000000	0.999988	0.000022	0.000026
30.40	0.000001	0.999989	0.000020	0.000025
30.50	0.000000	0.999989	0.000019	0.000023
30.60	0.000001	0.999990	0.000018	0.000022
30.70	0.000000	0.999991	0.000017	0.000021
30.80	0.000001	0.999991	0.000016	0.000020
30.90	0.000000	0.999992	0.000015	0.000019
31.00	0.000000	0.999992	0.000015	0.000018
31.10	0.000000	0.999992	0.000014	0.000017
31.20	0.000000	0.999993	0.000013	0.000016
31.30	0.000001	0.999993	0.000012	0.000015
31.40	0.000000	0.999994	0.000012	0.000014
31.50	0.000000	0.999994	0.000011	0.000013
31.60	0.000000	0.999994	0.000010	0.000013
31.70	0.000000	0.999995	0.000010	0.000012
31.80	0.000000	0.999995	0.000009	0.000011
31.90	0.000000	0.999995	0.000009	0.000011

TABLE 9 (cont.)

AMOUNT	FREQUENCY	CUMULATIVE	NET PREMIUM	EXP PREMIUM
32.00	0.000000	0.999995	0.000008	0.000010
32.10	0.000000	0.999996	0.000008	0.000010
32.20	0.000000	0.999996	0.000007	0.000009
32.30	0.000000	0.999996	0.000007	0.000009
32.40	0.000000	0.999996	0.000007	0.000008
32.50	0.000000	0.999996	0.000006	0.000008
32.60	0.000000	0.999997	0.000006	0.000007
32.70	0.000000	0.999997	0.000006	0.000007
32.80	0.000000	0.999997	0.000005	0.000006
32.90	0.000000	0.999997	0.000005	0.000006
33.00	0.000000	0.999997	0.000005	0.000006
33.10	0.000000	0.999998	0.000004	0.000005
33.20	0.000000	0.999998	0.000004	0.000005
33.30	0.000000	0.999998	0.000004	0.000005
33.40	0.000000	0.999998	0.000004	0.000005
33.50	0.000000	0.999998	0.000004	0.000004
33.60	0.000000	0.999998	0.000003	0.000004
33.70	0.000000	0.999998	0.000003	0.000004
33.80	0.000000	0.999998	0.000003	0.000004
33.90	0.000000	0.999998	0.000003	0.000003
34.00	0.000000	0.999998	0.000003	0.000003
34.10	0.000000	0.999999	0.000003	0.000003
34.20	0.000000	0.999999	0.000002	0.000003
34.30	0.000000	0.999999	0.000002	0.000003
34.40	0.000000	0.999999	0.000002	0.000003
34.50	0.000000	0.999999	0.000002	0.000002
34.60	0.000000	0.999999	0.000002	0.000002
34.70	0.000000	0.999999	0.000002	0.000002
34.80	0.000000	0.999999	0.000002	0.000002
34.90	0.000000	0.999999	0.000002	0.000002
35.00	0.000000	0.999999	0.000002	0.000002
35.10	0.000000	0.999999	0.000001	0.000002
35.20	0.000000	0.999999	0.000001	0.000002
35.30	0.000000	0.999999	0.000001	0.000002
35.40	0.000000	0.999999	0.000001	0.000001
35.50	0.000000	0.999999	0.000001	0.000001
35.60	0.000000	0.999999	0.000001	0.000001
35.70	0.000000	0.999999	0.000001	0.000001
35.80	0.000000	0.999999	0.000001	0.000001
35.90	0.0000000	0.999999	0.000001	0.000001
36.00	0.000000	1.000000	0.000001	0.000001

TABLE 10. *Net Premiums*

	$P(F^{\ell},\alpha)/P(F,\alpha)$		$P(F^{u},\alpha)/P(F,\alpha)$	
α	$d = 2$	$d = 1$	$d = 1$	$d = 2$
0	.924	1.000	1.00	1.000
1	.886	.983	1.000	1.002
2	.825	.972	1.000	1.000
3	.805	.952	1.000	1.025
4	.723	.955	1.002	1.004
5	.700	.931	1.005	1.045
6	.596	.902	1.008	1.023
7	.585	.905	1.019	1.085
8	.478	.863	1.008	1.041
9	.468	.852	1.017	1.114
10	.371	.830	1.020	1.076
11	.365	.802	1.029	1.162
12	.284	.794	1.038	1.126
13	.277	.760	1.032	1.212
14	.209	.741	1.046	1.170
15	.206	.718	1.044	1.275
16	.151	.696	1.058	1.226
17	.151	.677	1.063	1.361
18	.107	.655	1.068	1.288
19	.109	.633	1.081	1.451
20	.075	.611	1.079	1.358

TABLE 11. Exponential Premiums

	$P(F^{\ell},\alpha,0.1)/P(F,\alpha,0.1)$		$P(F^{u},\alpha,0.1)/P(F,\alpha,0.1)$	
α	$d = 2$	$d = 1$	$d = 1$	$d = 2$
0	.875	.981	1.003	1.012
1	.842	.969	1.004	1.016
2	.790	.959	1.005	1.017
3	.761	.942	1.006	1.035
4	.690	.939	1.007	1.026
5	.658	.918	1.010	1.056
6	.569	.893	1.014	1.046
7	.546	.889	1.023	1.094
8	.454	.854	1.016	1.069
9	.435	.839	1.024	1.127
10	.352	.818	1.027	1.106
11	.338	.792	1.036	1.176
12	.268	.781	1.044	1.157
13	.257	.750	1.040	1.229
14	.198	.731	1.053	1.205
15	.191	.709	1.053	1.293
16	.144	.687	1.066	1.264
17	.140	.668	1.071	1.379
18	.102	.646	1.076	1.329
19	.101	.625	1.089	1.468
20	.072	.604	1.089	1.404

REFERENCES

[1] Bühlmann, H. (1970). Mathematical Methods in Risk Theory. Springer, New York.

[2] Bühlmann, Gagliardi, Gerber, Straub (1975). "Some inequalities for stop-loss premiums," paper presented at the ASTIN-Colloquium in Portugal. Submitted to the ASTIN Bulletion.

[3] Gerber, H. (1974). "On iterative premium calculation principles," Swiss Actuarial Journal *74*, 163-172.

[4] Landingham, H. F., and Shariq, S. A. (1974). "Numerical evaluation of Compound Poisson distributions," American Inst. of Industrial Eng. Transactions, *6*, 50-54.

[5] Mereu, J. A. (1972). "An algorithm for computing expected stop-loss claims under a group life contract," TSA *24*, 311-320.

SIMULATION OF A MULTIRISK COLLECTIVE MODEL

John A. Beekman
Clinton P. Fuelling

Department of Mathematical Sciences
Ball State University
Muncie, Indiana

1. INTRODUCTION

Assume that $\{X_i\}$ is a sequence of independent, identically distributed random variables with a common distribution function $P(x)$. Assume that $E(X_i) = p_1 > 0$. Assume that $\{N(t), t \geq 0\}$ is a non-negative integral valued stochastic process, independent of the $\{X_i\}$ with $N(0) = 0$, and $E\{N(t)\} = t$.

Let $C(t) = \sum_{i=1}^{N(t)} X_i$, $0 \leq t < \infty$. This process describes the evolution of claim patterns, recognizing both the random number of claims $N(t)$, in time t, as well as the random nature of the claims, the X_i's. The time parameter is operational time which measures time by the expected number of claims during the period. (See page 36 of [3], for example.)

In [2] and [5], it was assumed that the $\{N(t), t \geq 0\}$ process was a Poisson stochastic process, in which case the $C(t)$ process is referred to as a compound Poisson process.

As in [2] and [5], let $S(t) = -I(t) + O(t) + L(t)$, $0 \leq t < \infty$

ISBN 0-12-394680-8

where the I(t), O(t), and L(t) processes are Ornstein-Uhlenbeck processes, independent of C(t) and each other, with zero mean functions and constant variance functions of σ_I^2, σ_O^2, σ_L^2 respectively. We will assume that

$$E\{I(s)I(t)\} = \sigma_I^2 e^{-\beta(t-s)} \quad \text{for} \quad s < t;$$

$$E\{O(s)O(t)\} = \sigma_O^2 e^{-\beta(t-s)} \quad \text{for} \quad s < t;$$

$$E\{L(s)L(t)\} = \sigma_L^2 e^{-\beta(t-s)} \quad \text{for} \quad s < t.$$

Reasons for choosing these processes to model investment performance deviations, operating expense deviations, and lapse expense deviations are given in [2]. Let $K^2 = \sigma_I^2 + \sigma_O^2 + \sigma_L^2$ for $K > 0$. By our assumptions,

$$f(s) = \frac{d}{ds} P[S(t) \leq s] = \{2\pi K^2\}^{-\frac{1}{2}} \exp\{-s^2/(2K^2)\}, \text{ for } -\infty<s<\infty.$$

Let $\rho = e^{-\beta(t-s)}$ and $A = [K^2(1 - \rho^2)]^{\frac{1}{2}}$. As explained in [2],

$$P[S(t) \leq y | S(s) = x] = \frac{1}{\sqrt{2\pi A^2}} \int_{-\infty}^{y} \exp-\left\{\frac{[w - x\rho]^2}{2A^2}\right\} dw.$$

The random variable $\theta = \dfrac{S(t) - x\rho}{A}$ has a N(0,1) distribution. The transition density function is

$$p(x,s;y,t) = \frac{\partial}{\partial y} P\{S(t) \leq y | S(s) = x\}$$

$$= [2\pi A^2]^{-\frac{1}{2}} \exp\left\{-\frac{[y - x\rho]^2}{2A^2}\right\}.$$

This conditional probability density function helps to describe the structure of the sample paths of the $S(\tau)$ process. Using the definition of S(t), the independence of the three processes, and previous assumptions, we see that $E\{S(s)S(t)\} = K^2 e^{-\beta(t-s)}$ for $s < t$. The correlation between $S(\tau)$ observations separated by one unit of time is $e^{-\beta}$. Observations of the $S(\tau)$ process can be used to estimate $e^{-\beta}$. Thus, $e^{-\beta}$ can be regarded as the theoretical autocorrelation function for lag $k = 1$ in equation (5) of

Miller-Hickman [11]. Equation (1) of that reference indicates how an observed series can be used to calculate the sample autocorrelation coefficient with lag 1:

$$r = \frac{\sum_{i=2}^{n} (Z_i - \overline{Z})(Z_{i-1} - \overline{Z})}{\sum_{i=1}^{n} (Z_i - \overline{Z})^2}$$

where n is the number of observations and

$$\overline{Z} = \frac{1}{n} \sum_{i=1}^{n} Z_i .$$

Let $R(t) = C(t) + S(t)$, $0 \leq t < \infty$. This stochastic process models the randomness of claims, and deviations from interest, lapse, and operating expense assumptions.

We will use Monte Carlo simulation of the sample paths and probabilities for the R(t) process. The simulation technique is largely based on techniques in [12] by H. L. Seal and [9] by I. I. Gringorten. Part of the method for the S(t) process is directly analogous to equation (4) of [11], which describes the autoregressive model of order 1.

Let Ω be the sample function space of the R(t) process. Each sample function w(t) of the C(t) process is determined as follows:

$$w(t) = 0, \quad 0 \leq t < t_1$$

$$w(t) = a_i, \quad \sum_{j=1}^{i} t_j \leq t < \sum_{j=1}^{i+1} t_j, \quad i = 1,2,\ldots$$

where $t_i > 0$ for $i = 1,2,\ldots$; $-\infty < a_i < \infty$ for $i = 1,2,\ldots$; $a_{i+1} - a_i \neq 0$ for $i = 1,2,\ldots$; and $\sum_{i=1}^{\infty} t_i = +\infty$. The t_i's represent the interoccurence times, and the a_i's represent the accumulated claims. The condition that $\sum_{i=1}^{\infty} t_i = +\infty$ allows at most a finite number of discontinuities in any finite interval. As explained on page 89 of [3], the sample functions of an Ornstein-

Uhlenbeck process are continuous. By Theorem 1 of [5], $S(t)$ is also an Ornstein-Uhlenbeck process. Thus, adding the $S(t)$ process to the $C(t)$ process can be visualized as perturbing the $C(t)$ step-like sample functions. Let λ be the aggregate safety loading for one operational time unit. Then the company charges $(p_1 + \lambda)t$ as the collective net premium for t operational time units. Denote the initial amount of funds assigned to the particular risk line by a non-negative number u.

Two sample paths are portrayed below:

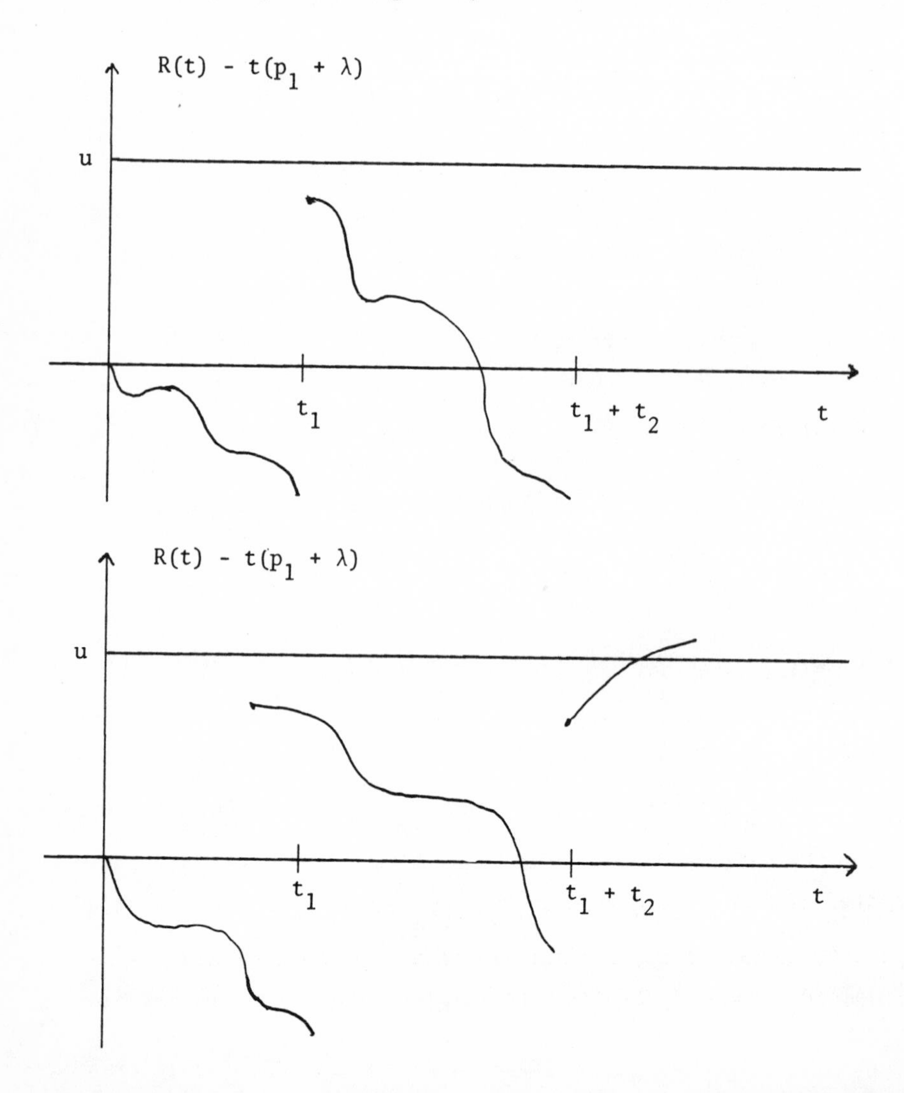

In the first case, the claims were exorbitant; in the second case, adverse deviations from investment, lapse, and/or operating expense assumptions exceeded previous expectations.

2. SIMULATION OF THE MODEL

Let

$$\tilde{\psi}(u,T) = P[\sup_{0 \le t \le T} \{C(t) + S(t) - t(p_1 + \lambda)\} > u].$$

We will now describe the steps which are followed in simulating the sample paths, and counting those paths which exceed the quantity u.

1. Select a suitable claim distribution P(x), and values for u, λ, and T. Load values of p_1, $\overline{C}$ (a division number used in later steps), K^2, $\rho = e^{-\beta}$, and A.

2. Compute a random variate, t_1, from the interclaim time distribution $P[T \le t]$, $t \ge 0$.

3. Divide the half-open interval $(0,t_1]$ into C parts by points $p_{1k} = \{kt_1/C\}$, $k = 1,2,\ldots,C$, where $C = \max\{2, [t_1 \cdot \overline{C}/E(T) + .8]\}$ and [x] is the greatest integer in x.

4. Draw a random normal variate, θ_1, from a N(0,1) distribution. Compute $S(p_{11}) = A\theta_1$, and $S(p_{11}) - p_{11}(p_1 + \lambda) = G(p_{11})$. If $G(p_{11}) > u$, record a "1" and start back at step 2 with a new risk enterprise. If $G(p_{11}) \le u$, compute $S(p_{12})$ as follows. Draw a second random normal variate θ_2. Compute $S(p_{12}) = \rho S(p_{11}) + A\theta_2$, and $G(p_{12}) = S(p_{12}) - p_{12}(p_1 + \lambda)$. If $G(p_{12}) > u$, record a "1" and start back at step 2 with a new enterprise. If $G(p_{12}) \le u$, compute $S(p_{13})$ as follows. Draw a third random normal variate θ_3. Compute $S(p_{13}) = \rho S(p_{12}) + A\theta_3$, and $G(p_{13}) = S(p_{13}) - p_{13}(p_1 + \lambda)$. Sequentially proceed through $k = C - 1$.

5. When $k = C$, draw a variate x_1 from P(x). Compute $H(t_1) = x_1 + G(t_1)$. If $H(t_1) > u$, record a "1" and start back at

step 2 with a new enterprise. If $H(t_1) \leq u$, select another variate t_2 from the interclaim time distribution.

6. Divide the half-open interval $(t_1, t_1 + t_2]$ into C parts by points $p_{2k} = \{t_1 + kt_2/C\}$, $k = 1,2,\ldots,C$, where $C = \max\{2, [t_2\overline{C}/E(T) + .8]\}$.

7. Using paragraph 4, sequentially check that $x_1 + S(p_{2k}) - p_{2k}(p_1 + \lambda) \leq u$ ($> u$) through $k = C - 1$.

8. When $k = C$, draw another variate x_2 from $P(x)$. Compute $H(t_1 + t_2) = x_1 + x_2 + G(t_1 + t_2)$. If $H(t_1 + t_2) > u$, record a "1" and start back at step 2 with a new enterprise. If $H(t_1 + t_2) \leq u$, select another variate t_3 from the interclaim time distribution.

9. If $t_1 + t_2 + t_3 > T$, the "company" has survived. A zero should be recorded. If $t_1 + t_2 + t_3 \leq T$, repeat steps 6, 7, 8, and 9 for new points $\{p_{3k}\}$, using $x_1 + x_2$ and $t_1 + t_2 + t_3$.

10. Continue until $t_1 + t_2 + \ldots + t_J > T$ for some J.

11. If n such financial histories (sample paths) are simulated, and k are ruined in the time $[0,T]$, then the estimate of $\tilde{\psi}(u,T)$ is $\hat{\psi}(u,T) = k/n$ with an estimated standard error of $\{(\frac{k}{n})(1 - \frac{k}{n})/n\}^{\frac{1}{2}}$. Print both numbers, along with earlier parameter values.

3. IMPLEMENTATION OF THE ALGORITHM

The simulation was carried out using a computer program written primarily in the FORTRAN language that was executed on an IBM 360 model 50 at the Ball State University Computer Center. The method for generating random deviates of a given distribution was accomplished by the procedure of Monte Carlo methods (see page 75 of [15]). Briefly, the procedure is to draw an ordinary random number and then use the cumulative distribution function of the given distribution to generate the desired random deviate.

The computer program consists of a main routine which controls the logic as given in paragraph 2. Separate subroutines were written for the generation of the random variables of the distributions listed in paragraph 4.

The ordinary random numbers were produced by a pseudorandom number generator that produces u_i from the uniform distribution in the open interval (0,1). The generator is based on the multiplicative congruential method of $y_{i+1} = 663608941\ y_i \pmod{2^{32}}$

$$u_i = y_i/2^{32}$$

as given in [1]. The random number seed, y_o, was taken as a small prime integer for each distribution. The pseudorandom number generator was written in assembler to take advantage of the IBM 360 architecture.

The routines for the negative exponential and the Pareto random deviates were written using the explicit inverses of the cumulative distribution functions (c.d.f.'s).

The routines for the random deviates with c.d.f.'s of

$$P[T \leq t] = 1 - 0.25e^{-0.4t} - 0.75e^{-2t} \quad \text{for } t \geq 0$$

and

$$P[X \leq x] = \sum_{i=1}^{5} a_i(1 - \exp(-b_i x)) \quad \text{for } x \geq 0$$

as used in examples 3, 4, and 5 in paragraph 4, found the inverse of the c.d.f. by the Newton-Raphson method. Two special considerations were required to use this iterative method. First, since the derivatives of the c.d.f.'s do not exist at 0, a negative iterate is adjusted to a positive value by setting its value equal to half of the previous iteration. Second, in order to keep the number of iterations small, a table of 100 initial values was produced at the beginning of the program execution. This table was produced for the probability values of .005,.015,..., .995. Thus given a random number r between 0 and 1, the initial iterate s_0 for the Newton-Raphson method can be found such that the value of the c.d.f. at s_0 is within .005 of the value of r. Using this procedure the method usually required 2 or 3 iterations.

The routine for the normal random number generation used the TR algorithm of Ahrens [1].

The tables in the next paragraph were produced by running the program on the IBM 360 model 50 computer using the FORTRAN H (code optimizing) compiler. The amount of central processor (CP) time to produce each table is as follows:

Table	CPU Time
1	9 hours 50 minutes
2	6 hours
3	6 hours 5 minutes
4	6 hours 15 minutes
5	7 hours 20 minutes

4. EXAMPLES

For our first example, we will use the negative exponential distribution for both the claim and interclaim time distributions. Thus

$$P(x) = \begin{cases} 1 - e^{-x}, & x \geq 0 \\ 0, & x < 0 \end{cases}$$

and

$$P[T \leq t] = \begin{cases} 1 - e^{-t}, & t \geq 0 \\ 0, & t < 0. \end{cases}$$

We will let T = 20, β = .5, $\overline{C}$ = 10, K = 0 or 1 or 2 and use selected values of u and λ. The results are displayed in Table 1. When K = 0, the S(t) process is suppressed and comparisons can be made with previous research on the C(t) process. The reader will observe that the values when K = 0 agree quite closely with those obtained by H. L. Seal in [12] for $\lambda = 0.05$ and 0.10, $u = 0$, 5, and 10 and with those obtained by Beekman and Bowers in [6] for $\lambda = 0.2$ and 0.3, $u = 0$, 5, and 10 and $\lambda = 0.1$, 0.2, and 0.3, $u = 15$, and 20.

As our second example, we let

$$P(x) = \begin{cases} 1 - (1 + 2x)^{-3/2}, & x \geq 0 \\ 0 & , x < 0 \end{cases}$$

and let $P[T \leq t]$ be as in example 1. The claim distribution is a Pareto distribution and is used by Seal in [12], Bohman in [7], and Thorin and Wikstad in Tables 5 and 6 of [14]. This is sometimes labelled a "dangerous distribution" as its variance is infinite. We considered four sets of parameter values. In each case, $\lambda = .3$, $\overline{C} = 10$, and $\beta = 0.5$. The parameters u, T, and K were allowed to vary as indicated below. The numbers of trials are indicated in Table 2.

For the third example, P(x) was retained as in the second example, but

$$P[T \leq t] = \begin{cases} 1 - 0.25e^{-0.4t} - 0.75e^{-2t}, & t \geq 0 \\ 0 & , t < 0. \end{cases}$$

Again $\lambda = .3$, $\overline{C} = 10$, and $\beta = 0.5$. The values obtained are displayed in Table 3.

For the fourth and fifth examples, $P(x) = \sum_{i=1}^{5} a_i[1-\exp(-b_i x)]$ for a_i's and b_i's given on page 148 of [14]. See also Table 7 of that reference. We let $\lambda = .1$, $\overline{C} = 10$, and $\beta = 0.5$, and $P[T \leq t]$ be as in Example 1 to create Table 4.

In the fifth example, $P[T \leq t]$ is as in the third example, and as in Table 8 of [14]. The λ, $\overline{C}$ and β values are as in the fourth example. The output is in Table 5.

Table 1

NO	P1	LAMBDA		SD(K)	RES(U)	TIME	N	RUIN	FREQNCY	EST-STD-ERR
1	1.0	0.05	0.606531	0.0	0	20	6000	5117	0.85283	0.45736E-02
2	1.0	0.05	0.606531	1.0	0	20	6000	5960	0.99333	0.10506E-02
3	1.0	0.05	0.606531	2.0	0	20	6000	6000	1.00000	0.0
4	1.0	0.05	0.606531	0.0	5	20	6000	1972	0.32867	0.60642E-02
5	1.0	0.05	0.606531	1.0	5	20	6000	2573	0.42883	0.63893E-02
6	1.0	0.05	0.606531	2.0	5	20	6000	3858	0.64300	0.61853E-02
7	1.0	0.05	0.606531	0.0	10	20	6000	586	0.09767	0.38325E-02
8	1.0	0.05	0.606531	1.0	10	20	6000	845	0.14083	0.44907E-02
9	1.0	0.05	0.606531	2.0	10	20	6000	1398	0.23300	0.54576E-02
10	1.0	0.05	0.606531	0.0	15	20	6000	127	0.02117	0.18583E-02
11	1.0	0.05	0.606531	1.0	15	20	6000	172	0.02867	0.21543E-02
12	1.0	0.05	0.606531	2.0	15	20	6000	334	0.05567	0.29600E-02
13	1.0	0.05	0.606531	0.0	20	20	6000	22	0.00367	0.78030E-03
14	1.0	0.05	0.606531	1.0	20	20	6000	38	0.00633	0.10241E-02
15	1.0	0.05	0.606531	2.0	20	20	6000	64	0.01067	0.13262E-02
16	1.0	0.10	0.606531	0.0	0	20	6000	5039	0.83983	0.47349E-02
17	1.0	0.10	0.606531	1.0	0	20	6000	5945	0.99083	0.12304E-02
18	1.0	0.10	0.606531	2.0	0	20	6000	5998	0.99967	0.23568E-03
19	1.0	0.10	0.606531	0.0	5	20	6000	1717	0.28617	0.58349E-02
20	1.0	0.10	0.606531	1.0	5	20	6000	2330	0.38833	0.62919E-02
21	1.0	0.10	0.606531	2.0	5	20	6000	3584	0.59733	0.63315E-02
22	1.0	0.10	0.606531	0.0	10	20	6000	513	0.08550	0.36099E-02
23	1.0	0.10	0.606531	1.0	10	20	6000	724	0.12067	0.42053E-02
24	1.0	0.10	0.606531	2.0	10	20	6000	1134	0.18900	0.50544E-02
25	1.0	0.10	0.606531	0.0	15	20	6000	117	0.01950	0.17851E-02
26	1.0	0.10	0.606531	1.0	15	20	6000	165	0.02750	0.21112E-02
27	1.0	0.10	0.606531	2.0	15	20	6000	294	0.04900	0.27868E-02
28	1.0	0.10	0.606531	0.0	20	20	6000	22	0.00367	0.78030E-03
29	1.0	0.10	0.606531	1.0	20	20	6000	32	0.00533	0.94029E-03

30	1.0	0.10	0.606531	2.0	20	20	6000	68	0.01133	0.13666E-02
31	1.0	0.20	0.606531	0.0	0	20	6000	4767	0.79450	0.52165E-02
32	1.0	0.20	0.606531	1.0	0	20	6000	5896	0.98267	0.16849E-02
33	1.0	0.20	0.606531	2.0	0	20	6000	5995	0.99917	0.37254E-03
34	1.0	0.20	0.606531	0.0	5	20	6000	1435	0.23917	0.55071E-02
35	1.0	0.20	0.606531	1.0	5	20	6000	1927	0.32117	0.60280E-02
36	1.0	0.20	0.606531	2.0	5	20	6000	3185	0.53083	0.64427E-02
37	1.0	0.20	0.606531	0.0	10	20	6000	349	0.05817	0.30217E-02
38	1.0	0.20	0.606531	1.0	10	20	6000	470	0.07833	0.34688E-02
39	1.0	0.20	0.606531	2.0	10	20	6000	871	0.14517	0.45478E-02
40	1.0	0.20	0.606531	0.0	15	20	6000	71	0.01183	0.13960E-02
41	1.0	0.20	0.606531	1.0	15	20	6000	86	0.01433	0.15345E-02
42	1.0	0.20	0.606531	2.0	15	20	6000	179	0.02983	0.21963E-02
43	1.0	0.20	0.606531	0.0	20	20	6000	9	0.00150	0.49962E-03
44	1.0	0.20	0.606531	1.0	20	20	6000	20	0.00333	0.74411E-03
45	1.0	0.20	0.606531	2.0	20	20	6000	29	0.00483	0.89536E-03
46	1.0	0.30	0.606531	0.0	0	20	6000	4463	0.74383	0.56354E-02
47	1.0	0.30	0.606531	1.0	0	20	6000	5856	0.97600	0.19759E-02
48	1.0	0.30	0.606531	2.0	0	20	6000	5989	0.99817	0.55227E-03
49	1.0	0.30	0.606531	0.0	5	20	6000	1103	0.18383	0.50006E-02
50	1.0	0.30	0.606531	1.0	5	20	6000	1538	0.25633	0.56366E-02
51	1.0	0.30	0.606531	2.0	5	20	6000	2729	0.45483	0.64286E-02
52	1.0	0.30	0.606531	0.0	10	20	6000	229	0.03817	0.24735E-02
53	1.0	0.30	0.606531	1.0	10	20	6000	342	0.05700	0.29931E-02
54	1.0	0.30	0.606531	2.0	10	20	6000	634	0.10567	0.39687E-02
55	1.0	0.30	0.606531	0.0	15	20	6000	34	0.00567	0.96907E-03
56	1.0	0.30	0.606531	1.0	15	20	6000	51	0.00850	0.11852E-02
57	1.0	0.30	0.606531	2.0	15	20	6000	131	0.02183	0.18866E-02
58	1.0	0.30	0.606531	0.0	20	20	6000	6	0.00100	0.40804E-03
59	1.0	0.30	0.606531	1.0	20	20	6000	9	0.00150	0.49962E-03
60	1.0	0.30	0.606531	2.0	20	20	6000	22	0.00367	0.78030E-03

TABLE 2

NO	P1	LAMBDA	RHO	SD(K)	RES(U)	TIME	N	RUIN	FREQNCY	EST-STD-ERR
1	1.0	0.30	0.606531	10.0	130	100	1200	27	0.02250	0.42811E-02
2	1.0	0.30	0.606531	100.0	1300	100	1200	2	0.00167	0.11775E-02
3	1.0	0.30	0.606531	10.0	130	1000	600	65	0.10833	0.12688E-01
4	1.0	0.30	0.606531	100.0	1300	1000	600	6	0.01000	0.40620E-02

TABLE 3

NO	P1	LAMBDA	RHO	SD(K)	RES(U)	TIME	N	RUIN	FREQNCY	EST-STD-ERR
1	1.0	0.30	0.606531	100.0	1300	100	1200	0	0.0	0.0
2	1.0	0.30	0.606531	10.0	130	100	1200	36	0.03000	0.49244E-02
3	1.0	0.30	0.606531	100.0	1300	1000	600	5	0.00833	0.37112E-02
4	1.0	0.30	0.606531	10.0	130	1000	600	83	0.13833	0.14095E-01

TABLE 4

NO	P1	LAMBDA	RHO	SD(K)	RES(U)	TIME	N	RUIN	FREQNCY	EST-STD-ERR
1	1.0	0.10	0.606531	100.0	1300	100	1200	0	0.0	0.0
2	1.0	0.10	0.606531	10.0	130	100	1200	33	0.02750	0.47209E-02
3	1.0	0.10	0.606531	100.0	1300	1000	600	8	0.01333	0.46825E-02
4	1.0	0.10	0.606531	10.0	130	1000	600	118	0.19667	0.16227E-01

TABLE 5

NO	P1	LAMBDA	RHO	SD(K)	RES(U)	TIME	N	RUIN	FREQNCY	EST-STD-ERR
1	1.0	0.10	0.606531	100.0	1300	100	1200	0	0.0	0.0
2	1.0	0.10	0.606531	10.0	130	100	1200	39	0.03250	0.51189E-02
3	1.0	0.10	0.606531	100.0	1300	1000	600	7	0.01167	0.43838E-02
4	1.0	0.10	0.606531	10.0	130	1000	600	134	0.22333	0.17003E-01

5. DISCUSSION OF OTHER REFERENCES

The reader would profit from a study of Harald Bohman's papers [7] and [8] on simulation techniques applied to risk problems, and Ove Lundberg's paper [10] related to [8].

H. L. Seal's paper [13] contains a wealth of material on the numerical calculation of survival probabilities for risk enterprises.

The author's papers [4] and [5] are concerned with fundamental results about the S(t) and R(t) processes.

REFERENCES

[1] Ahrens, J. H. and U. Dieter (1972). Computer Methods for Sampling from the Exponential and Normal Distributions, Communications of the ACM, *15*, 873-882.

[2] Beekman, John A. (1974). A New Collective Risk Model, Transactions of the Society of Actuaries, *25*, 573-597 (with discussion).

[3] _________ (1974). Two Stochastic Processes, Almqvist and Wiksell, Stockholm; also, Halsted Press (c/o John Wiley & Sons), New York.

[4] _________ (1975). Asymptotic Distributions for the Ornstein-Uhlenbeck Process, Journal of Applied Probability, *12*, 107-114.

[5] _________. Compound Poisson Processes, as Modified by Ornstein-Uhlenbeck Processes, Parts I, II, Scandinavian Actuarial Journal, 1975, 226-232, and 1976, 30-36.

[6] _________, and Newton L. Bowers, Jr. (1972). An Approximation to the Finite Time Ruin Function, Part II, Skandinavisk Aktuarietidskrift, 128-137.

[7] Bohman, Harald (1971). A Risk-Theoretical Model of Insurance Business, Skandinavisk Aktuarietidskrift, 50-52.

[8] _________ (1974). On the Use of Simulation Technique, Scandinavian Actuarial Journal, 175-178.

[9] Gringorten, Irving I. (1968). Estimating Finite-Time Maxima and Minima of a Stationary Gaussian Ornstein-Uhlenbeck Process by Monte Carlo Simulation, Journal of the American Statistical Association, *63*, 1517-1521.

[10] Lundberg, Ove (1974). Comments on Harald Bohman's Paper, Scandinavian Actuarial Journal, 179-181.

[11] Miller, Robert B. and James C. Hickman (1974). Time Series Analysis and Forecasting, Transactions of the Society of Actuaries, *25*, 267-329 (with discussions by Frank G. Reynolds and Richard W. Ziock).
[12] Seal, Hilary L. (1969). Simulation of the Ruin Potential of Nonlife Insurance Companies, Transactions of the Society of Actuaries, *21*, 563-585.
[13] __________ (1974). The Numerical Calculation of U(w,t), the Probability of Non-ruin in an Interval (0,t). Scandinavian Actuarial Journal, 121-139.
[14] Thorin, Olof and Nils Wikstad (1973). Numerical Evaluation of Ruin Probabilities for a Finite Period. ASTIN Bulletin, *7*, 137-153.
[15] Wooddy, John C. (1963). Study Note on Risk Theory, Society of Actuaries, Chicago.

EXPERIMENTAL COMPUTATION

Richard Vitale[1]

Division of Applied Mathematics
Brown University
Providence, Rhode Island 02912

1. INTRODUCTION

In recent years Professor Ulf Grenander has led an attempt here at Brown to integrate the computer into the study of mathematics, particularly in the areas of probability and statistics. In his earlier comments he described some of the activities which have been pursued in the course of teaching and research. At the risk of some overlap I should like to describe some of my own experiences and impressions with particular emphasis on the computer as an *experimental tool*. In large measure this attitude has been fostered by Professor Grenander's leadership and guidance.

I should like to begin by describing a mini-scenario. A mathematician sits at his desk engrossed with a theoretical conjecture. It seems plausible enough, but a rigorous proof is not

[1]*Present address: Department of Mathematics, Claremont Graduate School, Claremont, California 91711.*

ISBN 0-12-394680-8

yet at hand. He can, it seems, proceed in at least two ways. One is to tackle the problem in its fullest generality, marshalling powerful abstract concepts in order to solve the problem directly. A second way is to try to edge toward an understanding of the problem by simplifying it somehow or, more particularly, by looking at special cases. This latter approach, I think, can be fairly called *experimental mathematics:* gathering specific computations as a field geologist gathers rocks.

Unfortunately the mathematical literature often presents only final products -- elegant theorems combined in final language. The excitement of the search is often been drained in the presentation. In support of the experimental attitude I should like to present some quotations. The first is by D. H. Lehmer [3] and speaks of some of the "masters":

> If one examines the collected works of Euler, Gauss, Legendre, to mention but three, one finds them shamelessly and laboriously computing examples of empirical discoveries. Often their efforts led to the establishment of important theorems.

Hermann Weyl [5] put it this way once:

> But definite concrete problems were first conquered in their individual complexity, singlehanded by brute force, so to speak. Only afterwards the axiomaticians came along and stated: Instead of breaking in the door with all your might and bruising your hands, you should have constructed such and such a key of skill, and by it you would have been able to open the door quite smoothly. But they can construct the key only because they are able, after the breaking in was successful, to study the lock from within and without. Before you generalize, formalize, and axiomatize, there must be mathematical substance.

And, finally, Richard Courant [1]

> ...the flight into abstract generality must start from and return again to the concrete and specific.

Now let us return to our mathematician pondering his conjecture - let us say about matrices. The 2×2 case falls neatly into place. With a little more work the 3×3 case is proved and provides some additional insight. Undoubtedly the 4×4 case will yield another valuable clue, but unfortunately the computations are beastly. It is at this point, and historically natural, for the computer to enter. With a convenient system - at best a highly interactive one - experimentation can go on immediately.

Fortunately we here at Brown have had access to just such an opportunity and have been trying to exploit it as much as possible. Before going on to describe some of our specific activities, I should like to make another point. In a something of a turnabout, the accessibility of the computer has modified the kinds of mathematics we have been doing. The capabilities of the computer have in effect directed - and often re-directed - us to formulate different kinds of results. In particular we have been affected by the computer's pictorial output. Often a striking picture has changed the course of an investigation. Interestingly, Davis [3] has described some of the broader possibilities of a return to such "visual mathematics".

2. INSTRUCTION

Within instructional programs (both undergraduate and graduate) we have employed the computer as an aid to teaching through a research of "case method" approach. In dealing with such classical topics as limit theorems and regression, as well as more modern ones such as stochastic control, survey sampling, queuing, and search problems, an attempt is made to engage the student in discovering the underlying problems, formulating a solution, and testing the validity of the solution.

A typical project at the undergraduate level is simulation or, more precisely, Monte Carlo integration. Class time is first devoted to a discussion of the basic procedures together with motivation for their applications. The students are then assigned a number of experiments to perform in order to observe the behavior of the Monte Carlo estimator. Beyond this they are urged to carry out individual experiments and to keep a notebook describing interesting results. Their experiences are then discussed in class with the appropriate behavior usually (though not always!) reported. At this point theoretical results are presented concerning rates of convergence, computational efficiency, and so forth. The final phase of the project is devoted to variance reducing techniques such as stratification and the use of antithetic variables.

Another topic is stochastic approximation. Within a development similar in spirit to the above, experimentation and class discussion combine to explore fundamental behavior, theoretical results, and improvements (here accelerating methods).

Fortunately we have found that this approach engages most students right from the beginning. The possibility of doing their own "research" seems compelling. And we have had very few complaints of the form "Yes, those theorems are useful and the proofs are elegant but how did anyone get the original idea?" A second aspect is that the student may observe "spurious" behavior in his experimentation due to such effects as round-off or discretization problems, convergence-divergence ambiguity, or improbable random sequences. These problems often provide a convenient opportunity to discuss practical aspects of numerical computation.

Here is a sample of comments from an anonymous questionnaire filled out at the end of one course:

> "I particularly enjoyed the programming and found that it did improve my understanding of the material."

> "Computing results reinforced my understanding of the analysis presented in class."

> "Experimental stuff was really fun."

It is only fair to offer some of the criticisms too!

> "I was bored by the programming lectures."

> "One of the most interesting aspects of this course was the experimental attitude - but time (personal and computer) limitations induced a stodgy attitude. Do it the standard way or run out of computer time or skip experimenting (creative play) just when things get interesting."

3. RESEARCH

At the research level the computer has proved to be a useful ally in proving new results. I should like to describe two examples.

Estimation of the Mean of a Time Series[1]

Let us assume that $\{y_k\}$, $k = 1,2,\ldots$ is a real-valued, zero-mean discrete-time stationary time series with covariances

$$r_h = Ey_k y_{k\pm h} = \int_{-\pi}^{+\pi} e^{ih\lambda} f(\lambda)d\lambda$$

where the spectral density $f(\cdot)$ is continuous. Observations are of the form

$$x_k = m + y_k, \quad k = 1,\ldots,n,$$

and it is desired to estimate the unknown mean m. If we consider only linear, unbiased estimates of the form $\hat{m} = \sum_{k=1}^{n} c_k x_k$, $\sum_{k=1}^{n} c_k = 1$, it is possible to show that the minimum-variance unbiased estimate is given by

$$m_{BLU} = \sum_{k=1}^{n} c_k^{BLU} x_k$$

[1] *A full discussion can be found in Vitale* [4].

where the vector c^{BLU} is proportional to $R^{-1}\begin{bmatrix}1\\ \vdots\\ 1\end{bmatrix}$ (R the covariance matrix of $x_1,\ldots,x_n$). Such an estimate is of dubious practical importance since a matrix inversion, of possibly high order, is required. More critically, the covariance structure itself may be unknown. An attractive solution was devised by Grenander [2] who showed that the least-squares estimate $m_{LS} = \frac{1}{n}\sum_{k=1}^{n} x_k$ is asymptotically efficient in the sense that $Var(m_{BLU})/Var(m_{LS})$ approaches unity. This result holds, however, only under the restriction that $f(0) > 0$. At this point computer was brought in to investigate the case where $f(0) = 0$. Some initial work was done first to verify Grenander's results. First, we considered the efficiency of the least-squares estimate for $f(\lambda) = \frac{1}{2\pi}(1-.1 \cos \lambda)$ and $f(\lambda) = \frac{1}{2\pi}(1-.9 \cos \lambda)$ respectively. There was approach to unity in both examples, though slower in the second where $f(0)$ is significantly smaller. Satisfied with this verification, we proceeded to consider the case $f_0(\lambda) = \frac{1}{2\pi}(1-\cos \lambda)$ where $f(0) = 0$. Here we found apparent convergence to 0. This was sufficient evidence for abandoning the least-squares estimate as a candidate for asymptotic efficiency in the new context. To look for an alternative, it seemed reasonable to examine the actual coefficients in the minimum variance estimate. These were computed and plotted for the case $f_0(\lambda)$. The parabolic shape of the plot suggested that an estimate with $c_k = A + Bk + Ck^2$, quadratic in the index, might be asymptotically efficient. With this experimental evidence, it was then rather direct to prove the following result.

Theorem. If $f(\lambda)$ is continuous and positive except at $\lambda = 0$ where $\lim_{\lambda\to 0} \frac{f(\lambda)}{\lambda^2} \to 0$, then

$$\frac{Var(m_{BLU})}{Var(m_{LS})} \to 0 \qquad (1)$$

$$\frac{\mathrm{Var}(m_{BLU})}{\mathrm{Var}(m_{PAR})} \to 1 \tag{2}$$

where m_{PAR}, the "parabolic estimate", has

$$c_k = \frac{6n}{n^2-1}\left[\frac{k}{n}\left(1 - \frac{k}{n}\right)\right], \quad k = 1,\ldots,n.$$

Further, it was possible to show in the case $f(0) > 0$ that the parabolic estimate does not fare too badly:

$$\frac{\mathrm{Var}(m_{BLU})}{\mathrm{Var}(m_{PAR})} \to \frac{5}{6}.$$

A Problem From Pattern Analysis

Suppose that a function ϕ is chosen at random (this can be made precise) so that ϕ maps [0,1] onto [0,1] in such a way that

(i) ϕ is continuous

(ii) ϕ is strictly increasing; $\phi(0) = 0$, $\phi(1) = 1$.

It is possible to show that if ϕ_1 and ϕ_2 are chosen in this way then their composition $\psi = \phi_2 \circ \phi_1$ likewise satisfies (i) and (ii). It is of some interest to investigate what happens if this process is iterated. That is, if $\phi_1,\ldots,\phi_n$ are chosen independently, what does $\psi_n = \phi_n \circ \phi_{n-1} \circ \ldots \circ \phi_1$ look like? In the language of probability theory, is there a limiting distribution?

With the aid of the computer several cases were examined by simulation (note that in the previous study deterministic calculation rather than simulation was the tool). A typical situation was the so-called *fair* case in which $E\phi_i(x) = x$ for each x. Here many different models were studied and a number of the ψ_n were plotted. Invariably, and especially for large n, the pictures were similar. For values of $x <$ some α, ψ_n was very close to zero. At the critical values a precipitous increase was exhibited so that for $x > \alpha$, ψ_n was very close to one. What was of particular interest was that from simulation to simulation α

varied although within *any one* sequence it seemed to settle down. With this motivation it was then not hard to couch this behavior in more formal terms. In the fair case $\psi_n = \phi_n \circ \phi_{n-1} \circ \cdots \circ \phi_1$ converges (in an appropriate sense) to a function of the form

$$\psi_\infty(x) = \begin{cases} 0, & x < \alpha \\ 1, & x > \alpha \end{cases}$$

where α is uniformly distributed on [0,1].

REFERENCES

[1] R. Courant, Mathematics in the modern world, *Scientific American, vol. 211*, p. 43, 1964.

[2] Ulf Grenander, Statistical inference and stochastic processes, *Ark. Mat., vol. 1*, pp. 195-277, 1950.

[3] J. P. LaSalle (ed.). The Influence of Computing on Mathematical Research and Education, *Proc. Symp. in Appl. Math., vol. 20*, Am. Math. Soc., Providence, 1974.

[4] R. A. Vitale, An asymptotically efficient estimate in time series analysis, *Quart. Appl. Math., vol. 30*, pp. 421-440, 1973.

[5] H. Weyl, Emmy Noether, *Scripta Mathematics, vol. 3*, p. 214, 1935.

PARTITIONING FOR HOMOGENEITY

Donald E. McClure[1]

Division of Applied Mathematics
Brown University
Providence, Rhode Island 02912

SUMMARY

Several statistical problems are introduced that are interrelated by the common goal of decomposing a population into homogeneous parts. This goal leads to an optimal partitioning problem. Its analysis yields precise characterizations of optimal partitions of univariate and bivariate populations. The characterizations can be used for computation of approximately optimal partitions.

INTRODUCTION

Several distinct areas of statistics lead to a common problem of decomposing a population into a fixed finite number of homogeneous parts. The common mathematical problem has the following

[1]*This work was supported in part by the National Science Foundation, Grant MPS 75-01785.*

ISBN 0-12-394680-8

formulation. The population to be decomposed is represented by a probability density function f on a region Q in d-dimensional Euclidean space. Q will be partitioned into L subsets S_k, $k = 1,2,\ldots,L$. The *homogeneity* v of a single set S_k is measured by an average squared distance of the points in S_k from a fixed point a_k in Q:

$$v(S_k,a_k) = \frac{1}{w_k}\int_{S_k} ||x-a_k||^2 \, f(x)dx, \tag{1}$$

where

$$w_k = \int_{S_k} f(x)dx \tag{2}$$

In (1), $||x-a_k||$ denotes the Euclidean distance between x and a_k; w_k is the *weight* of the k-th subsets S_k. The *average homogeneity* $\mathscr{V}$ of the whole partition is measured by

$$\begin{aligned}\mathscr{V}(\mathscr{S}_L,\mathscr{A}_L) &= \sum_{k=1}^{L} w_k \, v(S_k,a_k) \\ &= \sum_{k=1}^{L} \int_{S_k} ||x-a_k||^2 \, f(x)dx;\end{aligned} \tag{3}$$

here $\mathscr{S}_L$ denotes the partition $\mathscr{S}_L = \{S_k\}_{k=1}^{L}$ and $\mathscr{A}_L$ denotes the point-system $\mathscr{A}_L = \{a_k\}_{k=1}^{L}$. The average homogeneity depends on the choice of $\mathscr{S}_L$ and $\mathscr{A}_L$. We are particularly interested in the extreme values of $\mathscr{V}$ and in the *optimal partitions* $\mathscr{S}_L^*$ and *optimal point-systems* $\mathscr{A}_L^*$ for which $\mathscr{V}$ assumes its minimum. We denote, for fixed L,

$$\mathscr{V}_L = \underset{\mathscr{S}_L,\mathscr{A}_L}{\text{infimum}} \; \mathscr{V}(\mathscr{S}_L,\mathscr{A}_L), \tag{4}$$

and we seek $\mathscr{S}_L^*$ and $\mathscr{A}_L^*$ such that

$$\mathscr{V}(\mathscr{S}_L^*,\mathscr{A}_L^*) = \mathscr{V}_L. \tag{5}$$

The functional $\mathscr{V}$ and its minimization are of interest in a number of diverse areas. In the next section we mention several applications to statistics. Then we shall review results which are relevant to the problem of computing approximations of optimal partitions and point-systems.

STATISTICAL APPLICATIONS

1. *Stratified Sampling*

The problem of optimal stratification of a univariate population for estimation of its mean value was considered by Dalenius [4]. Corresponding problems for multivariate populations have been well-formulated and analyzed by Isii and Taga [8]. Optimization of a stratified sampling scheme leads to the extremal problem of eqns. (3) - (5).

Suppose that the unknown mean vector μ of a d-variate population will be estimated from a sample of size n. The sample will be stratified; that is, the region Q where the population density f is positive will be partitioned into L *strata* S_k, $k = 1,2,\ldots,L$, and a specified portion of the total sample will be drawn at random from the k-th stratum. In so-called proportionate allocation, $n_k \cong n \cdot w_k$ observations are taken from S_k, where w_k is the weight (2) of S_k. The sample mean $\hat{\mu}_k$ of the observations from S_k is computed from the observations and is used to compute the estimate $\hat{\mu}$,

$$\hat{\mu} = \sum_{k=1}^{L} w_k \hat{\mu}_k,$$

of the population mean μ.

The estimator $\hat{\mu}$ is unbiased and its covariance matrix is

$$\mathscr{R}(\mathscr{S}_L) = \frac{1}{n} \sum_{k=1}^{L} \int_{S_k} (x-\mu_k)(x-\mu_k)^t \ f(x)dx. \tag{6}$$

Here $\mu_k = (1/w_k)\int_{S_k} x \ f(x)dx$ is the mean vector of the stratum S_k; the superscript "t" denotes vector transpose.

The covariance matrix depends on the choice of strata $\mathscr{S}_L$. A meaningful index of the quality of a particular choice of strata is the largest eigenvalue of $\mathscr{R}(\mathscr{S}_L)$. This eigenvalue corresponds to the variance of an estimator of a normalized linear functional of the mean vector μ. Specifically, it is the variance of $\eta^t\hat{\mu}$, an estimator of $\eta^t\mu$, where η is a unit eigenvector corresponding to the largest eigenvalue of $\mathscr{R}(\mathscr{S}_L)$.

For optimal stratification, we are then led to the problem of choosing $\mathscr{S}_L$ so that the largest eigenvalue of $\mathscr{R}(\mathscr{S}_L)$ is as small as possible. It can be shown that this objective is equivalent to the goal of minimizing the trace of $\mathscr{R}(\mathscr{S}_L)$. The trace of $n\mathscr{R}(\mathscr{S}_L)$ is given by the functional $\mathscr{V}$, eqn. (3), where the points a_k of the system $\mathscr{A}_L$ correspond to strata mean-vectors, $a_k = \mu_k$.

2. *Cluster Analysis*

Cluster analysis aims to identify distinct homogeneous subsets of a data set. Certain familiar and widely-used clustering procedures lead to the optimal partitioning problem of eqns. (3) - (5).

Let $\{X_j\}_{j=1}^n$ denote a set of n-points from a region Q of d-dimensional Euclidean space. The sample will be divided into L *clusters* S_k. We can regard the sets S_k as constituting a partition of Q. The *pooled within-group sum-of-squares* is commonly used to measure the collective homogeneity of the L clusters S_k. It is defined to be the value

$$V(\{X_j\}_{j=1}^n, \mathscr{S}_L) = \sum_{k=1}^{L} \sum_{X_j \in S_k} ||X_j - \hat{\mu}_k||^2 . \tag{7}$$

Here, as before, $\mathscr{S}_L$ denotes the partition $\{S_k\}_{k=1}^L$ and $\hat{\mu}_k$ is the average of those points X_j which are members of the k-th cluster.

If we can assume that the set $\{X_j\}_{j=1}^n$ is a random sample from a population with density f on Q, then we can relate the pooled within-group sum-of-squares (7) to the functional $\mathscr{V}$ of eqn. (3). Indeed, by the strong law of large numbers,

$$\frac{1}{n} V(\{X_j\}_{j=1}^n, \mathscr{S}_L) \rightarrow \mathscr{V}(\mathscr{S}_L, \mathscr{A}_L), \tag{8}$$

almost surely, as $n \rightarrow \infty$. In (8), $\mathscr{A}_L$ consists of the theoretical cluster means, $a_k = \mu_k = (1/w_k)\int_{S_k} x\, f(x)dx$. Thus, when n is large, the problem of clustering to minimize V, eqn. (7), is identified with the extremal problem for $\mathscr{V}$, eqns. (3) - (5).

3. *Data Compression*

One method of compressing a large sample from a continuous d-variate population is to discretize that population and then code each observation from the continuous population by replacing its value with a close representative from the approximating discrete set. A large sample from the continuous population is replaced in this process by a smaller set of tallies of the number of occurrences of the various elements in the discrete set. Let such an approximating discrete set have L members a_k and denote the set $\{a_k\}_{k=1}^L$ by $\mathscr{A}_L$. Let S_k (for $k = 1,2,\ldots,L$) denote the subset of the continuous population which will be associated with a_k through the discrete coding process. If the distribution of the continuous population is described by a probability density f, then the expected value of the squared-distance between a point chosen at random from the continuous population and its discrete representative will be given by the right side of eqn. (3). The problem of choosing a most representative discrete subpopulation $\mathscr{A}_L$ leads to the extremal problem of eqns. (3) - (5).

4. *Grouped Data*

An optimization problem in this setting is posed in a similar way to the one for data compression above, except that the problem for grouped data focuses on the choice of the sets S_k instead of the points a_k.

When collecting or analyzing sensitive data, one may conceal and protect individual observations by replacing point data with set data. For instance, one may ask a respondent to report which *interval*, among several possible ones, contains his or her personal income instead of reporting the exact value of that income. By such a procedure, one learns only which set of a partition of the sample space contains the sensitive datum.

In the general case of multivariate data, it is natural to identify all sample points in a particular set S_k with a point a_k in S_k. This parallels the common practice for grouped univariate data; see Cramér [3]. The functional $v(S_k,a_k)$ of eqn. (1) can be used to measure the error when sample points in S_k are associated with a_k. The average grouping error is then given by $\mathscr{V}$ of eqn. (3). The objective of "optimal grouping" leads again to the extremal problem for $\mathscr{V}$.

OPTIMAL PARTITIONS

The extremal partitioning problem has surfaced recurrently in the mathematical literature and in the literature of several areas of applications. An early paper by Steinhaus [10] contains the statement and derivation of several properties of optimal partitions. Steinhaus was motivated by a problem from the mechanics of rigid bodies.

We shall assume that Q, the domain of interest in R^d, is a convex set and that the density function f is positive on Q and continuous; these assumptions are made more for convenience than out of necessity.

Two basic properties tell us a great deal about the structure of optimal partitions and optimal point-systems:

(I) With any partition $\mathcal{S}_L = \{S_k\}_{k=1}^L$ given and fixed, it is necessary and sufficient for $\mathcal{A}_L = \{a_k\}_{k=1}^L$ to minimize $\mathcal{V}$ that $a_k = \mu_k$, for $k = 1,\ldots,L$, where

$$\mu_k = \frac{1}{w_k}\int_{S_k} x\ f(x)dx, \tag{9}$$

the *centroid* of S_k with respect to the density f.

(II) With any point-system $\mathcal{A}_L = \{a_k\}_{k=1}^L$ given and fixed, it is necessary and sufficient for $\mathcal{S}_L = \{S_k\}_{k=1}^L$ to minimize $\mathcal{V}$ that, for $k = 1,2,\ldots,L$,

$$S_k = \{x \in Q\colon ||x-a_k|| \leq ||x-a_j|| \text{ for all } j \neq k\}, \tag{10}$$

the *Dirichlet cell* of the point a_k in the system $\mathcal{A}_L$. (Eqn. (10) is necessary in a weak sense; S_k could differ from the right-side of (10) by a set of Lebesgue measure zero.)

Property (I) follows from the identity

$$\int_{S_k} ||x-a_k||^2 f(x)dx = \int_{S_k} ||x-\mu_k||^2 f(x)dx + w_k||\mu_k-a_k||^2.$$

Property (II) follows from the inequality

$$\sum_{k=1}^{L}\int_{S_k} ||x-a_k||^2 f(x)dx \geq \int_Q \min_{1\leq j\leq L} \{||x-a_j||^2\} f(x)dx;$$

the integrand on the right is always less than or equal to that on the left and equality is attained when each S_k satisfies eqn. (10).

Many authors have proposed the use of these two properties as the basis of an "algorithm" for computing optimal partitions. The algorithm applies these two properties alternately and iteratively. One specifies an initial point-system and partition; then the point-system is updated by computing centroids of the current partition; then the partition is updated by computing Dirichlet cells of the current point-system; and so on.

This algorithm is probably effective for the computation of

optimal univariate partitions. But when the dimension d is two or more, there are certain unavoidable complications. The most serious complication is nonuniqueness of partitions and point-systems satisfying properties (I) and (II) simultaneously. Examples are simple to construct when $d = 2$. Thus, the iterative algorithm above may not compute what it is designed to compute; the iteration need not converge to an optimal partition and an optimal point-system.

A second complication with this proposed method lies in the overwhelming amount of computation it entails. The computation of Dirichlet cells is a nontrivial task, especially when $d \geq 3$. It is probably not realistic to require this computation as an essential step in an iterative scheme.

Even though properties (I) and (II) may not solve all of the computational problems, they do give valuable insight into the structure of optimal partitions and point-systems. Optimal partitions are composed of convex polytopes (polygons when $d = 2$ and polyhedra when $d = 3$) intersected with Q. The individual convex polytopes are the Dirichlet cells of the associated optimal point-system. Optimal point-systems coincide with centroids of the sets in the associated optimal partition.

In one and two dimensions, we can obtain much sharper characterizations.

Optimal Univariate Partitions. The one-dimensional problem has received a substantial amount of attention in connection with its applications to stratified sampling designs. The analysis in this section is intimately related in spirit to the work of Dalenius and Hodges [5].

Suppose that Q is a bounded interval subset of the real line, $Q = [A,B]$, and that f is a positive and continuous probability density on Q. Let L be an arbitrary positive integer. From the preceding section, we know that an optimal partition will be composed of intervals. Let us restrict attention, therefore, to partitions $\mathcal{S}_L = \{S_k\}_{k=1}^L$ where $S_k = [t_{k-1}, t_k)$ and $A = t_o < t_1 < \ldots < t_L = B$. Let $\mathcal{A}_L$ be an arbitrary point-system in Q.

We shall obtain characterizations of optimal partitions by demonstrating a sharp lower bound for $\mathcal{V}(\mathcal{S}_L, \mathcal{A}_L)$ and by observing necessary conditions for attaining that bound. We let m_k denote the midpoint of $[t_{k-1}, t_k)$. The first step of the following expression uses the integral mean-value theorem; η_k is a point in S_k.

$$\begin{aligned}\mathcal{V}(\mathcal{S}_L, \mathcal{A}_L) &= \sum_{k=1}^{L} \int_{S_k} (x-a_k)^2 \, f(x)dx \\ &= \sum_{k=1}^{L} f(\eta_k) \int_{S_k} (x-a_k)^2 \, dx \\ &\geq \sum_{k=1}^{L} f(\eta_k) \int_{S_k} (x-m_k)^2 \, dx \\ &= \frac{1}{12} \sum_{k=1}^{L} f(\eta_k)(t_k - t_{k-1})^3 \\ &\geq \frac{1}{12} \frac{1}{L^2} \left[\sum_{k=1}^{L} f^{1/3}(\eta_k)(t_k - t_{k-1}) \right]^3. \end{aligned} \tag{11}$$

The last step invokes Hölder's inequality. A necessary and sufficient condition for attaining equality at the last step is that

$$f^{1/3}(\eta_k)(t_k - t_{k-1}) = \delta, \qquad \text{for } k = 1,2,\ldots,L, \tag{12}$$

where δ is a constant independent of k; see [7].

The sum in square brackets on the right side of (11) is a Riemann sum which approximates the integral of $f^{1/3}$. Indeed, by noting that equality can be attained in (11), under the condition of eqn. (12), we can deduce the asymptotic character of $\mathcal{V}_L$;

$$\lim_{L\to\infty} L^2 \, \mathcal{V}_L = \frac{1}{12}\left[\int_A^B f^{1/3}(x)dx\right]^3.$$

By summing both sides of (12), we can obtain an approximate equation for optimal partitioning points t_k;

$$\sum_{j=1}^{k} f^{1/3}(\eta_j)(t_j - t_{j-1}) = k \cdot \delta,$$

or

$$\int_A^{t_k} f^{1/3}(x)\,dx \cong \frac{k}{L} \int_A^B f^{1/3}(x)\,dx. \tag{13}$$

Equation (13) can be inverted to solve for t_k, $k = 1,2,\ldots,L$. The partition obtained in this way is suboptimal, but it is asymptotically equivalent to an optimal partition. It is reasonably straightforward to show that if $\mathcal{S}_L$ is computed from eqn. (13) and if $\mathcal{A}_L$ is the set of centroids of the intervals in $\mathcal{S}_L$, then

$$\lim_{L\to\infty} \mathcal{V}_L / \mathcal{V}(\mathcal{S}_L, \mathcal{A}_L) = 1;$$

in this sense, $\mathcal{S}_L$ and $\mathcal{A}_L$ are *asymptotically efficient*.

The condition (12) and the resulting equation for partitioning points (13) provide the means for explicit computation of good partitions. This approach may circumvent the complications of the iterative scheme mentioned above.

Optimal Bivariate Partitions. The two-dimensional partitioning problem has been studied extensively in connection with its applications to location theory. The two papers by Bollobás [1,2] contain recent contributions in this area.

In two dimensions it is possible to obtain precise characterizations, necessary conditions, for optimal partitions. These conditions can be used for the explicit construction of good approximations of optimal partitions. Such constructive methods were first described by L. Fejes Tóth [6] in work which was stimulated by the earlier paper of Steinhaus [10]. The present discussion is based on the recent paper by this author [9] in which strong and general necessary conditions for optimal partitions are proved. Since proofs are given in [9], we shall only give a brief description of the relevant results here.

Let Q be the unit square in R^2. (For the following results to hold, it suffices for Q to be Jordan measurable so that it can be approximated by a union of rectangles.) Suppose that f is a positive and continuous probability density function on Q. L is an arbitrary positive integer. We have noted that optimal partitions must be composed of convex polygons, so we shall restrict our attention to such partitions $\mathcal{S}_L$. (This permits the use of the integral mean-value theorem in a manner similar to its use for the univariate problem.) $\mathcal{A}_L$ denotes a point system $\{a_k\}_{k=1}^{L}$ in Q.

The analysis of the bivariate problem parallels that of the univariate problem; a lower bound for $\mathcal{V}(\mathcal{S}_L, \mathcal{A}_L)$ is first obtained and then necessary conditions for attaining that bound are observed. The two-dimensional problem involves an interesting geometric element; this makes the analysis more intricate.

The lower bound assumes the form

$$\mathcal{V}(\mathcal{S}_L, \mathcal{A}_L) \geq \frac{5\sqrt{3}}{54} \frac{1}{L} \left[\sum_{k=1}^{L} f^{1/2}(\eta_k) A_k \right]^2, \tag{14}$$

where A_k is the area of S_k and η_k is a point in S_k; η_k enters the expression when the mean-value theorem is used. The inequality (14) is the bivariate analog of (11). The following conditions are necessary for equality to hold in (14):

(i) for $k = 1,2,\ldots,L$, the set S_k is a regular polygon;

(ii) each set S_k is a hexagon;

(iii) for $k = 1,2,\ldots,L$,

$$f^{1/2}(\eta_k) A_k = \delta, \tag{15}$$

where δ is a constant independent of k.

The first two conditions are derived from extremal properties of regular polygons and from the fact that the average number of edges of the sets S_k in $\mathcal{S}_L$ cannot exceed six; this is a consequence of Euler's formula relating the numbers of vertices, edges, and faces of plane partitions. Condition (iii) and equation (15) are the bivariate analogs of (12).

It is not possible, unless Q has a very special structure, to satisfy all three conditions simultaneously. However the three conditions can be met by almost all sets S_k of a partition $\mathcal{S}_L$. For L sufficiently large, the proportion of the sets in $\mathcal{S}_L$ which will satisfy (i), (ii), and (iii) can be made arbitrarily close to one.

The sum in square brackets on the right side of (14) is a Riemann sum which approximates the integral of $f^{1/2}$. As we did for the univariate problem, we can deduce the asymptotic character of $\mathcal{V}_L$; in two dimensions,

$$\lim_{L\to\infty} L\cdot \mathcal{V}_L = \frac{5\sqrt{3}}{54}\left[\int_Q f^{1/2}(x)\,dx\right]^2.$$

The three necessary conditions above are the basis for methods of contructing good partitions. In some cases this can be done by conformal mapping of regular (and uniform) hexagonal partitions; an example is given in [9]. An alternative piecewise approximation method is suggested by L. Fejes Tóth [6]; it can be used when L is large. The idea is to divide the unit square Q into M subrectangles Q_j, for j = 1,2,...,M, where M is significantly smaller than L and where f is approximately constant on each subrectangle Q_j. The L sets in $\mathcal{S}_L$ are allocated to the separate subrectangles so that eqn. (15) will hold; the proportion allocated to Q_j is

$$\alpha_j = \int_{Q_j} f^{1/2}(x)\,dx \Big/ \int_Q f^{1/2}(x)\,dx.$$

Then each set Q_j is partitioned into (approximately) $L\alpha_j$ hexagons which are (approximately) regular and of identical area. If $\mathcal{S}_L$ is constructed in this way and if $\mathcal{A}_L$ is the set of centroids of the polygons in $\mathcal{S}_L$, then

$$\lim_{L\to\infty} \mathcal{V}_L / \mathcal{V}(\mathcal{S}_L, \mathcal{A}_L) = 1.$$

Thus, the construction yields *asymptotically efficient* partitions and point systems. The details of the construction and proofs of the results above are also given in [9].

REFERENCES

[1] B. Bollobás, The optimal structure of market areas, *J. Economic Theory 4*, pp. 174-179, 1972.

[2] B. Bollobás, The optimal arrangement of producers, *J. London Math. Soc. (2), 6*, pp. 605-613, 1973.

[3] H. Cramér, Mathematical Methods of Statistics, Princeton University Press, Princeton, 1946.

[4] T. Dalenius, The problem of optimum stratification, *Skand. Aktuarietidskr. 33*, pp. 203-213, 1950.

[5] T. Dalenius and J. L. Hodges, Jr., The choice of stratification points, *Skand. Aktuarietidskr. 40*, pp. 198-203, 1957.

[6] L. Fejes Tóth, Sur la représentation d'une population infinie par un nombre fini d'éléments, *Acta Math. Acad. Sci. Hungar. 10*, pp. 299-304, 1959.

[7] G. H. Hardy, J. E. Littlewood and G. Pólya, Inequalities, Second Edition, Cambridge University Press, London, 1952.

[8] K. Isii and Y. Taga, On optimal stratifications for multivariate distributions, *Skand. Aktuarietidskr. 52*, pp. 24-38, 1969.

[9] D. E. McClure, Characterization and approximation of optimal plane partitions, Report No. 36, Pattern Analysis Series, Division of Applied Mathematics, Brown University, Providence, Rhode Island, 1975.

[10] H. Steinhaus, Sur la division des corps matériels en parties, *Bull. Acad. Polon. Sci. Cl. III, 4*, pp. 801-804, 1956.

CORRELATES OF LIFE INSURANCE LAPSATION

J. R. Brzezinski

Life Insurance Management Research
Hartford, Connecticut

Life insurance persistency or lapsation has traditionally been measured in terms of an early lapse rate generally expressed as either (1) the proportion of business sold on which no part of the second year's premium is paid (the "traditional 13-month lapse rate") or (2) the proportion of business sold on which anything less than two full premiums is paid (the "traditional 24-month lapse rate"). Several years ago, the Life Insurance Marketing and Research Association "broke tradition" and also began to collect data on life insurance policy lapsation on all policies in force in a number of insurance companies and to investigate the results.

Last year, about this time, LIMRA published the first results of this project to study lapsation of all policies in a report to its membership "Long-Term Lapse Study." This report, as well as its sequel which should be distributed fairly soon, was more in the nature of various cross-tabulations of the data submitted from contributors rather than rigorous statistical testing.

The difficult task after publication of the tabulations was to determine and utilize various statistical and other techniques to convert the wealth of data that had been collected into

ISBN 0-12-394680-8

information about the nature of life insurance lapsation. Just the first year's contributions could be reduced to some 15,000 subtotals of data representing various classifications.

To make the statistical task somewhat more manageable and, hopefully, to make results somewhat more informative, it was decided to follow an approach toward investigating the data that somewhat parallels methods utilized by the Society of Actuaries in mortality investigations and to combine that with various statistical enrichments to recognize the various factors related to persistency.

The first step in approach was to develop a table of "expected lapse rates" to be used as a basis of comparison. The process was fairly standard from the actuarial point of view and resulted in the development of four sets of expected tables with separate tables by number and amount of insurance for pension business, high early cash value business, permanent insurance, and term insurance. Table 1 shows the values of the permanent expected rates. A simple regression test was utilized as a final check of the graduation process for the development of the expected tables and results came out very well -- a multiple correlation coefficient of .9, a constant term of .00006, and a regression coefficient of 1.0098 for the expected lapse rate -- a result pretty well confirming the expected tables. The results by amount of insurance were somewhat worse -- a multiple R of .825.

At this point, the investigation broke into two somewhat related but diverse investigations:

1. A further investigation of the variance and standard deviation of the difference between the actual and expected results was conducted. Our regression had indicated a standard error of estimate of .029. Although this seemed like an excellent estimate when dealing with a lapse in the first or second year, in later years, where an expected lapse rate of less than .02 could be anticipated the standard error was extremely high

relative to the expected value. It was felt that we were probably dealing with a situation in which the data is heteroscedastic and that it was not reasonable to assume that a uniform standard error throughout. A side benefit of this portion of the investigation would be that it would result in the development of a means of determining the nature of the variability of lapse rates and, possibly, give some indication where improvement in persistency is possible.

For this investigation, the summary data that was contributed for the first year of the study was subdivided into 100 categories. The subdivisions were designed as follows:

a. The data file was sorted by number of policies exposed in each summary card, and interval breakpoints were determined that would break the sample into 10 equally sized groupings.

b. Similarly, interval breakpoints were determined that would break the file into 10 equally sized groupings by expected lapse rate.

Within each of the 100 subcategories, the mean actual and expected lapse rates and the standard deviation of the difference between the actual and expected lapse rates were calculated. These data were then used to develop a "predictor equation" for the standard deviation of the difference between the actual and expected lapse rates. Table 2 illustrates the standard deviations observed in each "cell" and the resulting equation for the standard deviation. The most important predictor is the standard deviation that would be theoretically determined for a binomial distribution with expected q equal to the expected lapse rate. The other variables add considerably less prediction capability but tend to improve the prediction for selected values at the extremes of the table.

Although the results are not entirely unanticipated, having statistical confirmation and some quantitative method of estimating the standard deviation has a number of useful and practical applications:

Table 1. Graduated Select Lapse Rates - Permanent Life Insurance

Age	Policy Year														
	1	2	3	4	5	6	7	8	9	10	11	12	13	14	15
	Number of Policies														
0	.1120	.0556	.0399	.0337	.0321	.0258	.0231	.0199	.0171	.0151	.0127	.0125	.0124	.0121	.0117
1	.1512	.0619	.0433	.0364	.0335	.0255	.0220	.0191	.0172	.0157	.0127	.0124	.0121	.0117	.0119
2-4	.1911	.0681	.0413	.0332	.0316	.0254	.0228	.0199	.0176	.0162	.0141	.0139	.0143	.0156	.0187
5-9	.1912	.0701	.0389	.0295	.0281	.0227	.0210	.0192	.0183	.0188	.0197	.0248	.0294	.0332	.0382
10-14	.1439	.0539	.0304	.0234	.0253	.0244	.0283	.0292	.0290	.0301	.0328	.0342	.0353	.0362	.0379
15-19	.2153	.0843	.0471	.0391	.0461	.0459	.0446	.0407	.0373	.0359	.0360	.0357	.0344	.0325	.0303
20-24	.2718	.1168	.0662	.0530	.0519	.0441	.0404	.0358	.0319	.0294	.0276	.0265	.0253	.0237	.0219
25-29	.2264	.0931	.0556	.0457	.0445	.0371	.0336	.0295	.0263	.0243	.0228	.0222	.0215	.0202	.0182
30-34	.1972	.0792	.0494	.0421	.0419	.0351	.0320	.0280	.0245	.0224	.0206	.0201	.0197	.0190	.0177
35-39	.1732	.0697	.0472	.0421	.0415	.0334	.0288	.0244	.0215	.0204	.0192	.0190	.0181	.0170	.0171
40-44	.1440	.0582	.0422	.0386	.0384	.0311	.0274	.0237	.0214	.0205	.0194	.0190	.0180	.0168	.0169
45-49	.1280	.0520	.0410	.0382	.0377	.0303	.0268	.0238	.0222	.0220	.0217	.0219	.0217	.0218	.0248
50-54	.1140	.0468	.0401	.0384	.0385	.0316	.0288	.0260	.0249	.0256	.0288	.0314	.0316	.0309	.0313
55-59	.1012	.0459	.0420	.0406	.0413	.0373	.0364	.0342	.0333	.0328	.0304	.0272	.0245	.0227	.0239
60-64	.0837	.0659	.0568	.0486	.0435	.0311	.0263	.0238	.0225	.0212	.0190	.0173	.0160	.0158	.0177
65-69	.0816	.0507	.0448	.0391	.0357	.0277	.0246	.0216	.0189	.0171	.0149	.0151	.0170	.0211	.0276
70-99	.1095	.0819	.0644	.0497	.0393	.0278	.0242	.0222	.0205	.0188	.0160	.0143	.0121	.0096	.0070

Table 1 (cont.)

	Policy Year														
	1	2	3	4	5	6	7	8	9	10	11	12	13	14	15
Age	Amount of Insurance														
0	.0995	.0506	.0425	.0356	.0331	.0236	.0225	.0207	.0186	.0167	.0162	.0150	.0143	.0139	.0140
1	.1597	.0564	.0425	.0358	.0331	.0225	.0207	.0186	.0167	.0151	.0150	.0143	.0139	.0140	.0150
2-4	.1694	.0605	.0450	.0369	.0327	.0204	.0178	.0156	.0136	.0118	.0117	.0109	.0115	.0137	.0176
5-9	.1427	.0529	.0409	.0332	.0299	.0175	.0169	.0171	.0176	.0187	.0208	.0249	.0291	.0337	.0390
10-14	.1097	.0337	.0331	.0286	.0273	.0158	.0196	.0231	.0257	.0275	.0291	.0290	.0290	.0289	.0288
15-19	.2123	.0855	.0543	.0423	.0397	.0352	.0351	.0350	.0349	.0339	.0318	.0293	.0269	.0248	.0232
20-24	.2606	.1057	.0629	.0492	.0435	.0385	.0366	.0347	.0324	.0296	.0273	.0248	.0227	.0211	.0199
25-29	.1915	.0789	.0545	.0451	.0410	.0347	.0321	.0195	.0267	.0242	.0226	.0207	.0191	.0172	.0155
30-34	.1454	.0621	.0503	.0440	.0411	.0349	.0320	.0289	.0259	.0230	.0214	.0197	.0186	.0179	.0175
35-39	.1233	.0561	.0494	.0440	.0410	.0344	.0316	.0289	.0262	.0239	.0224	.0210	.0201	.0200	.0199
40-44	.0986	.0450	.0444	.0406	.0388	.0330	.0319	.0306	.0291	.0271	.0253	.0232	.0215	.0203	.0194
45-49	.0915	.0448	.0454	.0413	.0387	.0315	.0297	.0288	.0288	.0288	.0290	.0289	.0280	.0270	.0264
50-54	.1010	.0493	.0471	.0422	.0393	.0319	.0296	.0282	.0278	.0288	.0330	.0353	.0350	.0334	.0308
55-59	.0782	.0414	.0451	.0422	.0414	.0411	.0422	.0410	.0389	.0357	.0298	.0250	.0218	.0192	.0166
60-64	.0645	.0400	.0463	.0438	.0420	.0350	.0319	.0312	.0312	.0311	.0300	.0278	.0247	.0214	.0179
65-69	.0608	.0329	.0418	.0406	.0402	.0362	.0354	.0342	.0337	.0339	.0346	.0356	.0365	.0374	.0379
70-99	.0631	.0428	.0503	.0498	.0491	.0505	.0495	.0486	.0479	.0477	.0468	.0471	.0474	.0477	.0479

TABLE 2. Standard Deviation of Difference from Expected Value

Expected Rate Group	Policies Exposed per Record Grouping									
	1	2	3	4	5	6	7	8	9	10
1	0.05430	0.10709	0.05940	0.03066	0.02150	0.01793	0.01086	0.00967	0.00635	0.00461
2	0.17513	0.14188	0.09365	0.04135	0.02557	0.01791	0.01851	0.01005	0.00653	0.00614
3	0.18690	0.08238	0.08656	0.04121	0.02961	0.02957	0.01605	0.01294	0.01067	0.00658
4	0.16443	0.11611	0.06080	0.04724	0.03329	0.02055	0.01958	0.01511	0.01067	0.00901
5	0.14990	0.14350	0.12514	0.05994	0.04319	0.03451	0.02705	0.02200	0.01337	0.00937
6	0.24994	0.21032	0.10541	0.06309	0.05016	0.03667	0.03063	0.02265	0.03974	0.01089
7	0.27777	0.18011	0.15586	0.07727	0.05930	0.05376	0.04533	0.04285	0.06145	0.01610
8	0.33370	0.16490	0.13561	0.10358	0.08166	0.06789	0.04596	0.03388	0.02997	0.02870
9	0.27476	0.21526	0.17624	0.13969	0.07389	0.05986	0.06660	0.04280	0.04517	0.03506
10	0.37579	0.26162	0.19886	0.12609	0.10618	0.09029	0.07517	0.06314	0.06013	0.05288

Predicted Standard Deviation:

-.011 + 1.109 (expected binomial std. dev.) + .141(square root of expected lapse rate)
- .600(expected lapse rate ÷ number of policies exposed)

a. The applications in Risk Theory are self-evident. These results indicate somewhat the restraints that can be expected in applying the concept of a "lapse rate delta" in "released from risk" reserves.

b. The results can be utilized to "flag" unusual deviations in lapse rate studies.

c. The results could be used to create families of lapse rates that would reflect high or low lapse rate assumptions.

2. Concurrently with the investigation of the standard deviation, it was also decided to look into the degree to which the prediction of a lapse rate could be improved by considering additional characteristics of the business being studied. For this purpose, a variation of "dummy variable multiple regression" was chosen as the statistical technique to utilize. The "dummy variables" chosen were: Male, Female, Medical, Nonmedical, Medical and Policy Year 1, Nonmedical and Policy Year 1, Permanent Insurance, Pension Insurance, and High Early Cash Value Insurance. In addition to the dummy variables, two continuous variables were chosen: the Expected Lapse Rate (as previously calculated) and the natural logarithm of the average policy size. The regression model chosen was to fit an equation of the form $q = A \times {}_qE + B$ where A and B are column matrices of coefficients. To develop a means of getting the A coefficients, it was determined that these would result from utilizing independent variables equal to the expected lapse rate times each of the dummy variables. We called such variables "multiplicative variables."

Although the expected tables supposedly reflected differences associated with High Early Cash Values, Pension, Permanent, and Term, these variables were included in the regressions as it was anticipated that as relationships associated with other variables were developed, that adjustments would be needed with these variables to sort of take up the slack.

Table 3 shows the various simple correlation of each of the variables used in the regression with the actual lapse rate. Normally, these would probably give some good indication of the relative importance of the various variables, but it does appear as if the use of the dummy variable clouds issues somewhat in that correlations seem to be affected to some degree by the relative proportion of the sample in which a particular dummy is observed.

There are a number of problems with the use of this model and method of statistical analysis that researchers should keep in mind if they are to do it. We at LIMRA utilized a stepwise multiple regression program that had been modified to use summary data and to "weight" observations by the policy exposure. With such a program, the researcher simply cannot ignore the limitations on the computer actually being used.

We found considerable difficulty in reconciling results of runs to previous tabulations and regression runs. The difficulty was almost entirely attributable to the number of significant digits of calculation retained internally in the computer and the order in which input went into the machine (mostly the latter). Best results were obtained by presorting input into ascending order by exposure to maximize the retention of significant digits with internal floating point notation.

To a large extent, the difficulties with having heteroscedastic data are reduced by having the multiplicative variable in the regression. Such variables usually entered into the regression before their straight dummy variable counterpart. Now that we have the formula for the standard deviation, we will experiment with changing the weighting formulas in the multiple regression so that greater weight will be given to observations that have smaller standard deviations to offset the effect of overweighting some observations and having a faulty cutoff criterion.

Table 4 shows the results of the multiple regression with the additional variables. The additional variables raise the multiple

TABLE 3. Correlation of Variables to Actual Lapse Rate

	Variable	Simple Correlation Coefficient
1	Male (D)	.0192
2	Female (D)	-.0047
3	Medical (D)	-.1181
4	Non-Medical (D)	.1144
5	Medical -- year 1 (D)	.2558
6	Non-Medical -- year 1 (D)	.7059
7	Permanent Insurance (D)	-.1088
8	Pension Insurance (D)	.2740
9	High Early Cash Value Insurance (D)	.0081
10	Expected Lapse Rate (C)	.9032
11	Male (M)	.7047
12	Female (M)	.3176
13	Medical (M)	.2236
14	Non-Medical (M)	.7424
15	Medical -- year 1 (M)	.2701
16	Non-Medical -- year 1 (M)	.7277
17	Permanent Insurance (M)	.7043
18	Pension Insurance (M)	.2968
19	High Early Cash Value Insurance (M)	.0310
20	Ln Ave. Size Policy (C)	.0955

Key:

D = "Dummy Variable" (variable = 1 if condition is true otherwise variable = D)

C = "Continuous Variable

M = "Multiplicative Variable" (variable equals corresponding "Dummy Variable" times Expected Lapse Rate)

TABLE 4. Results of Multiple Regressions

Standard Error of Estimate: .0269

Multiple Correlation Coefficient: .917

Constant Term: .119

	Variable	Coefficient	Beta Coefficient
1	Male (D)	-.0089	-.0578
2	Female (D)	-.0155	-.0955
3	Medical (D)	.0091	.0662
4	Non-Medical (D)	.0095	.0687
5	Medical -- year 1 (D)	.0411	.1172
6	Non-Medical -- year 1 (D)	.0295	.1123
7	Permanent Insurance (D)	-.0028	-.0151
8	Pension Insurance (D)	-.0013	-.0051
9	High Early Cash Value Insurance (D)	.0115	.0177
10	Expected Lapse Rate (C)	.7298	.6528
11	Male (M)	.2948	.2557
12	Female (M)	.2476	.1405
13	Medical (M)	-.0136	-.0085
14	Non-Medical (M)	.2570	.2292
15	Medical -- year 1 (M)	-.2863	-.1453
16	Non-Medical -- year 1 (M)	-.2043	-.1754
17	Permanent Insurance (M)	-.0817	-.0705
18	Pension Insurance (M)	-.0339	-.0170
19	High Early Cash Value Insurance (M)	.0211	.0025
20	Ln Average Size Policy (C)	-.0119	-.1375

correlation coefficient from .9 to .92 or reduce the unexplained variance reamining after considering the expected tables by nearly 20 per cent.

We at LIMRA have found these results useful in several ways. In response to the characteristic pressures of LIMRA's women's liberationists, we have developed a formula that does indicate that women do indeed have better persistency than their male counterparts considering all the variables in the regression. We have additional proof of the new expected tables and some general guidelines of what we can expect to find in tabulations of the continuing lapse study.

The greatest use we will have, however, is that we can utilize equations of the sort developed to follow trends over time and, with the upcoming expansion of the underlying study to include more variables (finer breakdowns of term insurance; mode of premium payment; policy loan status; size designations; agent classification; expansion to substandard, group conversions, term conversions, and guaranteed insurability option policies; and breakdown by country) and more companies contributing we have a method that we can use with confidence to evaluate the various correlates of life insurance lapsation.

SOME PRACTICAL NOTES ABOUT SOLVING THE LUNDBERG RISK EQUATION

Nils Wikstad

Försäkringstekniska Forskningsnämnden
Stockholm, Sweden

In order to calculate the probability of ruin in the classical ruin theory, there is a method developed by Olof Thorin involving the roots of the Lundberg risk equation. In general this equation is complex and in precomputer time it was prohibitively tedious to find solutions to it. With the aid of a computer it is not too difficult to solve the equation, but one must be aware that a method which has been shown to be fool-proof in the theory is not necessarily so in practice, mostly due to the limited number of significant figures the computer deals with.

In general the Lundberg risk equation is written:

$$f(s) = 1 - k(z - cs)\cdot p(s)$$

For the aim of this paper we just need to know that

k = the m.g.f. of interclaim times
p = the m.g.f. of claims.

We assume z to be a complex number and c a positive real number, in fact, the loaded unit-premium.

Furthermore we assume that the analysis has given the number of roots of the equation.

ISBN 0-12-394680-8

If the number of roots is unknown, it is impossible to be sure that all roots are found.

Two main choices have to be made:

1. Choose an iterative method, which converges to the root.
2. Choose an initial approximation.

There are many well-known methods, which can be thought of. Among them some are more attractive than others. The very first is the classical which rearranges the equation

$f(s) = 0$ into $s = g(s)$ e.g. when applied

$$s_{n+1} = g(s_n). \tag{1}$$

The method is near at hand in the case of exponential interclaim time d.f. is $(1 - e^{-t})$. This means that the m.g.f. $k(u) = \frac{1}{1-u}$ and as a consequence the Lundberg risk equation becomes

$$s = \frac{1}{c}[z - 1 + p(s)]$$

Now there are at least two drawbacks to this method. Let r be the root, then $|g'(r)|$ must be less than 1 for convergence and, if it converges, it converges slowly.

But in many cases there is more than one way to rearrange $f(s)$ into $s = g(s)$, so the condition $|g'(r)| < 1$ can be fulfilled. Let us assume that the convergence is slow: The method works even if the initial approximation is distant from the root.

A method worth trying secondly is Newton-Raphson's generalized by Fröberg.

$$s_{n+1} = s_n - \frac{f(s_n)}{f'(s_n)} - \frac{1}{2} \cdot \frac{f''(s_n)}{f'(s_n)} \cdot \left(\frac{f(s_n)}{f'(s_n)}\right)^2 - \cdots \tag{2}$$

This has two main drawbacks. First the initial approximation has to be close to the root, $|f'(r)|$ must not be zero, and $|f'(s)|$ must not be very small compared to $|f(s)|$ in the neighborhood of the root. You can add more terms, but it means a lot more calculations and you cannot be sure that you get a better result without examination of the remainder.

If f(s) is a polynomial there are many methods. Among them there is one which has been proved successful. It is developed by Dejon-Nickel. The main idea is to find a correction by examining each term in a Taylor-development of the polynomial but it would carry too far here to describe the details.

When the d.f. of claims and the d.f. of interclaim times both are of the exponential type

$$\sum_{\nu=1}^{n} a_\nu (1 - e^{-\alpha_\nu \cdot t}),$$

the m.g.f.'s are of the type $\sum \frac{a_\nu}{1 - \frac{s}{\alpha_\nu}}$. This means that the Lundberg equation can be written as a fractional polynomial, but then the arithmetical backlash is worsened in a not negligible way. Therefore it may be recommended, after evaluating a root with Nickel's method, to use the root as an initial approximation in a Newton-Raphson attack on the original equation.

This leads us over to the second problem, which often can be more difficult than expected. According to the author's experience, it is always possible to find an initial approximation which makes the chosen method to breakdown.

A well-known example is that of a real polynomial with complex roots. If you apply Newton-Raphson with a real initial approximation you will never find the complex roots.

In general, the origin is a self-evident initial approximation in most cases. One good reason for this is that in the case of a polynomial it is better to search for the absolutely smallest root in the first place, then eliminate the root by syntetic division, and after that look for the next smallest one. The division will accumulate round off errors which easily drowns a small root.

Before skipping a method one ought to have tried at least one initial approximation in each quadrant, because an axis can be an insuperable barrier for the actual method.

If there is some information from analysis or experience, *use it*!

At last, when a root is evaluated - at least you think so - test it in the original equation, not only in the one rearranged for the search! When looking for more than one root, test that the new root is not a copy of one already evaluated.

REFERENCES

Fröberg, C. E., "Introduction to Numerical Analysis," Addison-Wesley, 1965, pp. 23-24.

Dejon, B. and Nickel, K., "A Never Failing, Fast Convergent Root-Finding Algorithm," Constructive Aspects of the Fundamental Theorem of Algebra, Wiley-Interscience, London 1968, pp. 1-35.

Thorin, O., ASTIN Bulletin VI, 1 and 2.

Thorin, O. and Wikstad, N., "Numerical Evaluation of Ruin Probabilities for a Finite Period," ASTIN Bulletin VII, pp. 137-153.